Tourism and Agriculture

Shifting global consumption patterns, tastes and attitudes towards food, leisure, travel and place have opened new opportunities for rural producers in the form of agritourism, ecotourism, wine, food and rural tourism and specialized niche market agricultural production for tourism. Agriculture is one of the oldest and most basic parts of the global economy, whereas tourism is one of the newest and most rapidly spreading. In the face of current problems of climate change, rising food prices, poverty and a global financial crisis, linkages between agriculture and tourism may provide the basis for new solutions in many countries. A number of challenges, nevertheless, confront the realization of synergies between tourism and agriculture.

Tourism and Agriculture examines region-specific cases at the interface between tourism and agriculture, looking at the impacts of rural restructuring, and new geographies of consumption and production. To meet the need for a more comprehensive appreciation of the relationships and interactions between the tourism and agricultural economic sectors, this book considers the factors that influence the nature of these relationships and explores avenues for facilitating synergistic relationships between tourism and agriculture. These relationships are examined in 13 chapters through case studies from eastern and western Europe, Japan and the United States and from the developing countries of the Pacific, the Caribbean, Ghana and Mexico. Themes of diversification, economic development and emerging new forms of production and consumption are integrated throughout the entire book.

This essential volume built on original research generates new insights into the relationships between tourism and agriculture and future economic rural development. Edited by leading researchers and academics in the field, this book will be of value to students, researchers and academics interested in tourism, agriculture and rural development.

Rebecca Maria Torres is an Assistant Professor in the Department of Geography and the Environment at the University of Texas, Austin.

Janet Henshall Momsen is an Emerita Professor of Geography at the University of California, Davis. She is also a Senior Research Associate in the International Gender Studies Institute and in the Centre for Geography and Environmental Sciences at the University of Oxford.

Contemporary Geographies of Leisure, Tourism and Mobility

Series edited by C. Michael Hall

Professor at the Department of Management, College of Business & Economics, University of Canterbury, Christchurch, New Zealand

The aim of this series is to explore and communicate the intersections and relationships between leisure, tourism and human mobility within the social sciences.

It will incorporate both traditional and new perspectives on leisure and tourism from contemporary geography, for example, notions of identity, representation and culture, while also providing for perspectives from cognate areas such as anthropology, cultural studies, gastronomy and food studies, marketing, policy studies and political economy, regional and urban planning and sociology, within the development of an integrated field of leisure and tourism studies.

Also, increasingly, tourism and leisure are regarded as steps in a continuum of human mobility. Inclusion of mobility in the series offers the prospect to examine the relationship between tourism and migration, the sojourner, educational travel and second home and retirement travel phenomena.

The series comprises two strands:

Contemporary Geographies of Leisure, Tourism and Mobility aims to address the needs of students and academics, and the titles will be published in hardback and paperback. It includes the following titles:

The Moralisation of Tourism
Sun, sand . . . and saving the world?
Jim Butcher

The Ethics of Tourism Development
Mick Smith and Rosaleen Duffy

Tourism in the Caribbean
Trends, development, prospects
Edited by David Timothy Duval

Qualitative Research in Tourism
Ontologies, epistemologies and methodologies
Edited by Jenny Phillimore and Lisa Goodson

The Media and the Tourist Imagination
Converging cultures
Edited by David Crouch, Rhona Jackson and Felix Thompson

Tourism and Global Environmental Change
Ecological, social, economic and political interrelationships
Edited by Stefan Gössling and C. Michael Hall

Cultural Heritage of Tourism in the Developing World
Edited by Dallen J. Timothy and Gyan Nyaupane

Understanding and Managing Tourism Impacts
Michael Hall and Alan Lew

Forthcoming:
An Introduction to Visual Research Methods in Tourism
Edited by Tijana Rakic and Donna Chambers

Routledge Studies in Contemporary Geographies of Leisure, Tourism and Mobility is a forum for innovative new research intended for research students and academics, and the titles will be available in hardback only. It includes the following titles:

Living with Tourism
Negotiating identities in a Turkish village
Hazel Tucker

Tourism, Diasporas and Space
Edited by Tim Coles and Dallen J. Timothy

Tourism and Postcolonialism
Contested discourses, identities and representations
Edited by C. Michael Hall and Hazel Tucker

Tourism, Religion and Spiritual Journeys
Edited by Dallen J. Timothy and Daniel H. Olsen

China's Outbound Tourism
Wolfgang Georg Arlt

Tourism, Power and Space
Edited by Andrew Church and Tim Coles

Tourism, Ethnic Diversity and the City
Edited by Jan Rath

Ecotourism, NGOs and Development
A critical analysis
Jim Butcher

Tourism and the Consumption of Wildlife
Hunting, shooting and sport fishing
Edited by Brent Lovelock

Tourism, Creativity and Development
Edited by Greg Richards and Julie Wilson

Tourism at the Grassroots
Edited by John Connell and Barbara Rugendyke

Tourism and Innovation
Michael Hall and Allan Williams

World Tourism Cities
Developing tourism off the beaten track
Edited by Robert Maitland and Peter Newman

Tourism and National Parks
International perspectives on development, histories and change
Edited by Warwick Frost and C. Michael Hall

Tourism, Performance and the Everyday
Consuming the Orient
Michael Haldrup and Jonas Larsen

Tourism and Change in Polar Regions
Edited by C. Michael Hall and Jarkko Saarinen

Fieldwork in Tourism
Methods, issues and reflections
Edited by C. Michael Hall

Tourism and India
Kevin Hannam and Anya Diekmann

Political Economy of Tourism
Edited by Jan Mosedale

Volunteer Tourism
Edited by Angela Benson

The Study of Tourism
Richard Sharpley

Children's and Families' Holiday Experiences
Neil Carr

Tourism and Agriculture
Edited by Rebecca Maria Torres and Janet Henshall Momsen

Forthcoming:

Tourism and National Identity
Edited by Elspeth Frew and Leanne White

Gender and Tourism
Cara Atchinson

Tourism in China
David Airey and King Chong

Real Tourism
Edited by Claudio Minca and Tim Oaks

Tourism and Animal Ethics
David A. Fennell

Tourism and Agriculture

New geographies of consumption, production and rural restructuring

Edited by

Rebecca Maria Torres and Janet Henshall Momsen

Routledge
Taylor & Francis Group
LONDON AND NEW YORK

First published 2011
by Routledge
2 Park Square, Milton Park, Abingdon, Oxon OX14 4RN

Simultaneously published in the USA and Canada
by Routledge
711 Third Avenue, New York, NY 10017, USA

Routledge is an imprint of the Taylor & Francis Group, an informa business

Typeset in Times New Roman by
Book Now Ltd, London

British Library Cataloging in Publication Data
A catalogue record for this book is available from the British Library

Library of Congress Cataloging in Publication Data
A catalogue record has been requested for this book

ISBN: 978–0–415–58429–6 (hbk)
ISBN: 978–0–203–83440–4 (ebk)

Contents

Illustrations

Tables

Figures

Contributors

Alex Boakye Asiedu is an Associate Professor at the Department of Geography and Resource Development of the University of Ghana, Legon, Accra, Ghana. He is currently the chairman of the University of Ghana's Volta Basin Research Project. He has published widely in a number of renowned journals such as *International Journal of Tourism Research*, *Population, Space and Place* and *Journal of Housing and Built Environment*. His research interests are in the areas of tourism, international migration and urban housing development.

Tracy Berno has lived and worked in the South Pacific region for more than 20 years and was Head of Department – Tourism and Hospitality at the University of the South Pacific for many years. As a tourism academic and consultant, Tracy's main area of interest has been the relationship between local agricultural production and the tourism and hospitality industry in the South Pacific region. This interest grew into a research programme focusing on how the produce and cuisines of the South Pacific could be highlighted as part of the tourism product, and in doing so, how these linkages could be used as part of integrated rural development and sustainable livelihoods in the region. Tracy has researched and published in this area, and most recently, co-authored a book on the food and cuisine of the South Pacific. She is now the Planning Manager at Lincoln University, New Zealand, but continues with her consulting interest in the Pacific.

Gemma Canoves is a Professor of Geography at the Autonomous University of Barcelona, Spain and Director of Research and International Relations of Tourism Studies at the Autonomous University of Barcelona. She is the author or editor of 10 books and has published widely in a number of renowned journals such as *Annals of Tourism Research*, *Geoforum*, *Sociologia Ruralis* and *Tourism Geographies*. Her research interests are in areas of rural tourism and agriculture, gender, new tourisms and local development.

Deborah Che is an Affiliate Assistant Professor and Senior Fellow in the School of Public Policy at the George Mason University. Her research interests include sustainable economic development, natural resource-based tourism (i.e. agritourism, ecotourism and hunting) development and marketing, cultural/heritage tourism and arts-based economic diversification strategies. She has published in geography, tourism and social science journals including the *Annals of the Association of American Geographers*, *The Professional Geographer*, *Geoforum*, *Tourism Geographies*, *Tourism Recreation Research*, *Tourism Review International*, *Social Science Quarterly* and *Agriculture and Human Values*. She is on the editorial board of Tourism Geographies and has served as Chair of the Recreation, Tourism and Sport specialty group of the Association of American Geographers.

Rosie Cox is a Senior Lecturer in Geography and Gender Studies at Birkbeck, University of London, UK. She has a long-standing interest in consumption within the home and its intersection with inequality. She is the author of *The Servant Problem: Domestic Employment in a Global Economy* (2006, I.B. Tauris), co-author of *Reconnecting Consumers, Producers and Food: Exploring 'Alternative' Networks* (2008, Berg) and co-editor of *Dirt: New Geographies of Cleanliness and Contamination* (2007, I.B. Tauris).

Jill Donaldson is a co-founder of the Berkeley, CA-based Food Literacy and Nutrition Project, which aims to educate urban folk about food systems. During the past decade, she has contributed her expertise in community development and nutrition to a wide variety of programmes and organizations in the Bay Area. She holds an MS in Community Development from the University of California, Davis, where her research focused on agritourism.

Elizabeth Dowler works on food and poverty; local initiatives and policy evaluation; food security, rights and justice and 'reconnection' to sustainable food systems, especially consumers' perspectives. A Registered Public Health Nutritionist, she is Professor of food and social policy in the Department of Sociology, University of Warwick; she is also a member/trustee of the Food Ethics Council, an independent research and advocacy group working to make the food system fairer and healthier, and of Defra's Council of Food Policy Advisers.

Tometi K. Gbedema is a candidate for PhD in Geography at the University of California, Davis. He is currently working on his dissertation: 'The Gate of No Return – Role of Heritage Tourism in Local Communities in Sub-Saharan Africa: The Cases of Elmina and Keta in Ghana'. His book, *NIMBY: Natural Resource Development Issues* is currently at press with LAP Publishing in Germany. Gbedema's research interests are in geography, tourism, community and economic development in West Africa.

Atsuko Hashimoto is an Associate Professor in the Department of Tourism and Environment at the Brock University. Her research focuses on socio-cultural and human aspects (especially women and children) of tourism development. Selected publications include 'The Challenges of Developing Sustainable

Partnerships in Rural Tourism in Traditional Japan' in press 2010 with D.J. Telfer for *Journal of Policy Research in Tourism, Leisure and Events*.

Lewis Holloway is a Lecturer in Geography at the University of Hull, UK. His research interests cover agricultural technologies, alternative rural lifestyles and alternative food networks. With the co-authors of the chapter appearing in this volume, he recently published *Reconnecting Consumers, Producers and Food: Exploring 'Alternative' Networks* (2008, Berg).

Moya Kneafsey is a Reader in Human Geography at Coventry University. She researches the potential for 'alternative' food networks to deliver positive social, economic and environmental effects. She also maintains an active interest in rural and community tourism.

Irén Szörényiné Kukorelli is a Professor of Geography at the Széchenyi István University in Győr, Hungary. She is a rural social geographer, who has 25 years of experience at a social research institute, the Centre for Regional Studies of the Hungarian Academy of Sciences. Her interests are rural change, rural tourism and European rural development policy. She is the author of three books and has edited eight books. She has published 29 papers in Hungarian and foreign periodicals and 45 book chapters. Since 2009, she has been president of the Euracademy Association, a pan-European organization for sustainable rural development.

Janet Momsen is an Emerita Professor of Geography at the University of California, Davis and Senior Research Associate in the Centre for the Environment and International Gender Studies at the University of Oxford. She is the author or editor of 15 books and has taught in universities in Canada, the United States, Brazil, the Netherlands and the United Kingdom. Her research interests include gender and development, rural tourism and agriculture.

Stern Neill is an Associate Professor at the Orfalea College of Business at California Polytechnic State University. His scholarly interests are in the areas of marketing strategy, decision making and firm performance. He received his PhD in Business Administration from the Louisiana State University in 2000. His research appears in the *Journal of Business Research*, *Industrial Marketing Management* and others.

Pamela Richardson-Ngwenya is a Postdoctoral Fellow at the School of Development Studies, University of KwaZulu-Natal, South Africa. Her doctoral thesis was completed at the University of Oxford, entitled 'The Ethical Geographies of Caribbean Sugar'. Her research interests include video methodologies, feminist geographies, food and agriculture, environment and development.

Kevon Rhiney has recently submitted his doctoral thesis on agritourism linkages in Jamaica and currently lectures in the Department of Geography and Geology at the University of the West Indies, Mona. His specific research interests include agritourism linkages and rural sustainable development, farmers' associations and rural capacity building and urban and regional planning.

Ellen L. Rilla (Ellie) has worked in Marin County as the University of California (UC) Farm Advisor and Director of the Ag Extension office for the past 21 years. Her CV reflects a diversity of both academic and pragmatic work focused on land use policies and programmes about diversification, farm succession and long-term sustainability of the working landscapes of Northern California. In 1997, she spent several months in England and the East Coast immersing herself in the study of agritourism. Her explorations resulted in her co-authoring a UC publication, *Agritourism and Nature Tourism in California*. It has sold over 10,000 copies and she is currently completing the update of the second edition.

Raul Suhett de Morais is a PhD candidate at the Department of Geography of the Autonomous University of Barcelona, Spain, a researcher at the Tudistar Group, Spain and a consultant for Raízes Turismo, Brazil. He is currently working on his dissertation 'Tourist Image Evolution Through Guidebooks – The Case of Barcelona (Spain)'. He started publishing not long ago, but he has already published and presented papers at journals and academic events. His research interests are in tourism, geography, tourist image, city marketing and guidebooks.

David J. Telfer is an Associate Professor in the Department of Tourism and Environment at Brock University. His research focuses on tourism and development theory and the linkages between tourism and agriculture. Selected publications include *Tourism and Development in the Developing World* with R. Sharpley (Routledge, 2008).

Benjamin F. Timms is an Assistant Professor of Geography in the Social Sciences Department at the California Polytechnic State University. He studies sustainable tourism development in the Caribbean with a focus on maximizing economic linkages between tourism and local industries. His publications have appeared in *Global Development Studies*, *International Development Planning Review*, *Caribbean Geography* and others.

Rebecca Maria Torres is an Assistant Professor in the Department of Geography and the Environment at the University of Texas, Austin. She holds a PhD from the University of California, Davis, where she studied the linkages between tourism and agriculture in the Mexican state of Quintana Roo. Her current research reflects a range of interests linked to rural development and poverty reduction in Latin America and the US South. Currently, she examines migration, agricultural development and tourism in developing country economies in the context of globalization. She is presently engaged in a 5-year research, education and outreach project on transnational Latino migration and rural transformation in the US South supported by the National Science Foundation.

Laura Venn has a PhD from Manchester Metropolitan University and worked for 3 years as the Research Assistant on the AHRC/ESRC-funded project 'Reconnecting Consumers, Producers and Food: Exploring "alternative" networks' at Coventry University. She now works as a cultural research analyst in the public sector.

Astrid Wargenau is the Sales Director for ITB Asia for the Messe Berlin group of companies. She received an MA in geography from Western Michigan University and published her thesis research in the *International Journal of Wine Marketing*. In addition to her work on wine tourism, she has researched German-themed tourism in the United States. Her professional interests include tourism marketing, conference management and meeting coordination.

Acknowledgements

The editors wish to thank all the contributors for sharing their original research in this volume and for their diligence in completing the multiple iterations requested of them. We are extremely grateful to our research assistants, Ingrid Haeckel and Lindsey Carte, who worked tirelessly communicating with authors, copy editing, formatting and making tables, maps and graphs among other tasks. In addition, they provided insightful comments and feedback that helped improve the overall quality of the book. Also, we thank the editors and staff at Routledge who provided invaluable guidance throughout the process and worked with great timeliness and efficiency to produce this volume.

1 Introduction

Rebecca Maria Torres and Janet Momsen

In this volume, we bring together two sectors of the world economy and consider their interactions and linkages. Agriculture is one of the oldest and most basic sectors of the global economy, whereas tourism is one of the newest and most rapidly spreading. In the face of current problems of climate change, rising food prices and a global financial crisis, linkages between agriculture and tourism may provide the basis for new solutions in many countries. A number of challenges, nevertheless, confront the realization of synergies between tourism and agriculture. One prerequisite is enhanced understanding of the processes of rural restructuring and agricultural transformation and their association with tourism development – including better recognition of their potential role as beneficial, detrimental or disruptive. These relationships are examined in 13 chapters through case studies from eastern and western Europe, Japan and the United States and from the developing countries of the Pacific, the Caribbean and Ghana and Mexico.

Since the postwar era, global economic restructuring has been characterized by a declining agricultural sector and the rise of service-based economies. Although agriculture remains an important livelihood strategy for many rural people across the globe, tourism has become the world's largest and fastest growing industry. Harnessing its massive potential has become a key economic development strategy in both industrialized and developing nations. Until recently, the relationships between global tourism and local agriculture had not been explored in-depth. Recent research suggests that tourism and agriculture relationships tend to be multifaceted, place-specific and highly complex. The potential for creating synergistic relationships between tourism and agriculture has been widely recognized by development planners, policy makers and academics alike. However, realizing the benefit of those synergies has proven far more difficult than anticipated. Furthermore, there is growing evidence that the two sectors often vie for resources and evolve in a manner that is detrimental to local agriculture and rural communities because of competition for labour, land, water, investment and other resources. Demand for labour may lead to out migration from the countryside to tourist resorts, thus leaving underutilized farmland in the countryside and encouraging the growth of urban slums because of the lack of housing for workers (Torres and Momsen 2005; Chapter 4).

This book aims to fill a gap in the current literature by presenting regionally specific cases of the interface between tourism and agriculture, examining the impacts of rural restructuring and analyzing the implications of new geographies of consumption and production. Indeed, the book highlights a variety of emerging original research conducted on this topic that has neither previously been brought together nor conceptualized in this manner. To meet this need for a more comprehensive appreciation of the relationships and interactions between the tourism and agricultural economic sectors, this book will (1) examine the multiple relationships that exist between tourism and agriculture in various contexts, (2) consider the factors that influence the nature of these relationships and (3) explore avenues for facilitating and realizing the benefit of synergistic relationships between tourism and agriculture. Given these objectives, we have brought together 13 contributions from 23 scholars working in geography, history, agriculture, psychology, business, tourism studies and community and regional development. Given the place-specific nature of tourism and agriculture relationships, these contributions offer case studies from a wide range of geographic contexts. Although themes of diversification, economic development and emerging new forms of production and consumption are threaded throughout the entire book, the chapters fall roughly into three parts: (I) tourism, agriculture and rural restructuring; (II) building tourism and agricultural linkages: challenges and potential and (III) new forms of tourism and agricultural production and consumption.

Part I (Chapters 2–5) incorporates chapters examining the intersections between tourism and agriculture – both beneficial and detrimental to communities – and the implications of these interactions for rural restructuring, poverty alleviation, economic development and empowerment of rural communities and people. In their totality, the chapters illustrate the complex intersections between local and global forces in producing existing tourism and agricultural relations, processes of rural restructuring and contingencies that foster or constrain synergies.

Part II (Chapters 6–9) addresses the relationships between tourism and agriculture, with an emphasis on the challenges and potential for creating linkages between the two sectors to foster food production for tourism markets and agricultural development, contribute to regional development, alleviate poverty, decrease leakages through imports, improve tourism industry food supplies, increase tourist access to local foods, improve the gastronomic experience for visitors, foster unique regionalized food identities and reduce food miles. Increasing backward linkages between tourism and agriculture contributes to sustainability in both the tourism and agriculture sectors.

In Part III (Chapters 10–14), the authors explore the various new forms of tourism and agriculture production and consumption that are emerging as farmers diversify farm production and tourists continue to seek out new experiences, destinations and types of consumption in terms of food, place, ethics, leisure and entertainment. These new forms of tourism – including rural, pro-poor, sustainable, responsible, farm, gastronomic, wine and agritourism – have emerged as a growing niche market that provides local rural producers with alternative outlets and sources of income, while enhancing the tourism industry offerings. The demand for these new products is driven, in part, by global shifts in consumer tastes, preferences and ethics that increasingly favour local production and fair trade products, organic or natural foods, sustainable tourism and agriculture and desire to reconnect to farming, food, place and rurality. Indeed, both tourism and agricultural developments stand to gain through the synergies inherent in these new forms of production and consumption (Part III).

Shifting global consumption patterns, tastes and attitudes towards food, leisure, travel and place have opened up new opportunities for producers in the form of rural tourism, agritourism, ecotourism, wine tourism, food tourism and specialized niche market agricultural production for tourist consumption. Tourists allow farmers to reduce food miles by bringing the market to the farm and so cut the costs of distributing products.

For many rural dwellers, both farm and non-farm, the provision of both local produce and new leisure spaces for tourist consumption provides the only way for a viable lifestyle. Both Rilla and Che and Wargenau (Chapters 12 and 13, respectively) suggest that, because of the diversification and increased income associated with rural or agritourism, some family farms are able to maintain land in agriculture as opposed to selling for development or other uses, employing tourism income to enhance and expand farming operations.

Therefore, fomenting the creation of linkages between tourism and agriculture has recently received considerable attention as a strategy for rural and agricultural development in stagnating rural areas. As tourism and agriculture increasingly intersect and transform, there is growing interest among governments, the private sector, academics, donor agencies and non-profit organizations to better understand the relationship between these two sectors, to encourage interaction and to become involved in fostering these linkages. Chapter 3, on rural tourism in Ghana, illustrates the importance of the private sector and return migrants, whereas Chapter 9, on Barbados, provides an example of government and international agencies encouraging tourism's links with local agriculture. Similarly, Chapter 10 discusses the role of government in providing training, regulation and certification for rural and agritourism in Italy.

Creating linkages between tourism and agriculture, through production for tourism markets and/or agritourism, both necessitates and facilitates the opportunity for unique forms of marketing place and consumption. These new forms of marketing capitalize on new global consumption patterns manifest in the demand for high-quality, healthy, authentic, sustainable and responsible tourism and food products. For example, Berno (Chapter 6) discusses the emergence of 'farm-to-fork' concepts promoting the use of local agricultural products as part of the tourism experience. In one of the most novel approaches, Cox *et al.* (Chapter 10) discuss Italy's highly successful 'adopt-a-sheep' programme, where 'parents' are encouraged to visit their adoptees on the farm. Similarly, in the case of wine tourism, consumers are brought to the production space, which creates opportunities for education, quality assurance and place attachments associated with consuming products, as in the case of wine tourism described by Che and Wargenau

(Chapter 13) and in farm-stay tourism in California (Chapter 11). These strategies are premised on creating connections and relationships between consumers and producers, often based on bringing the tourist to the production place to consume.

Tourism and agriculture interact in many different ways. In this book, we explore many of these links through tourism, agritourism, agro-tourism, ecotourism, nature/green tourism, farm stays, tourist food supplies, sustainable tourism, food festivals and museums and alternative food networks. In many countries, these links are relatively new and related to political transformation and rural restructuring as in Hungary (Chapter 2), Japan (Chapter 5), Mexico (Chapter 4) and Spain (Chapter 14) and involve policy and planning by local and national governments. This is particularly seen in Europe where LEADER projects funded by the European Union have been used to gain local government support of rural tourism as in Spain (Chapter 14) and Hungary (Chapter 2). Cox *et al.* in Chapter 10 note that the European Union has also stimulated rural tourism through 'protected designation of origin' or a 'protected geographical indication'– appellations that identify and protect products from certain geographical origins or derived from artisanal methods. The initiatives for these changes may come from individual actions at the local level and from central government, or in the case of the European Union, pan-national-driven policies. In the chapters of this book, we examine tourism and agriculture relationships initiated at multiple scales in many different parts of the world.

Tourism has often been triggered by political change: as in Mexico with the adoption of neoliberal policies, in Spain with the end of fascism and in Hungary following the fall of Communism. In the latter two cases, the subsequent membership in the European Union led to freer movement of people. Rural tourism came later when the traditional tourist sites became crowded (beaches in Spain and Lake Balaton and Budapest in Hungary, Chapters 14 and 2, respectively). More recently, the search for an unspoiled environment as in the national parks of Hungary and England (Chapters 2 and 12, respectively) and in rural Spain became a major new attraction. Recent and growing interest in local food has further attracted tourists, especially in Europe, to rural areas as in Italy (Chapter 10). Interest in wine-growing areas has also attracted a special type of gourmet tourist as in California (Chapter 11), Michigan (Chapter 13) and Spain (Chapter 14), with tourism becoming a new form of marketing for wine producers. Japan has also started to see its rural areas as sites for tourism, in the form of farm stays, farmers' markets and farm museums, as the number of growers declines, their average ages rise and rice production falls (Chapter 5).

Changes in global trade patterns with the loss of subsidized prices for traditional plantation export crops following new World Trade Organization rulings have forced many small tropical islands in the Pacific and Caribbean to turn to tourism as a new source of foreign exchange (Scheyvens and Momsen 2008a). Chapters 7 and 8 suggest that restructuring of the global food system (including the food crisis) provides an impetus for the islands to break free of the legacy of plantation economies through increased domestic production for greater food security and to reduce leakages. Thus, linkages between local agriculture and tourism are becoming ever more important. Former sugar and banana farmers are learning how to grow vegetables, salad crops and fruit for tourist hotels. Hotels are also being weaned from importing foods from the nearest metropolitan countries such as the United States for the Caribbean (Chapters 7–9) and Australia and New Zealand for the western Pacific (Chapter 6). There are, however, various factors that influence the ability of farmers to produce for and access tourism markets, including the nationality, size and structure of hotels (all-inclusive versus standard); the nationality and training of chefs; the business experience of rural entrepreneurs and the quality and reliability of local produce (Chapters 2, 3 and 6–9). The chapters in Part II of this volume are built upon a significant literature on tourism and agriculture linkages (for a review see Torres and Momsen 2004) to expand this scope of research integrating issues of global restructuring of agro-food systems, neoliberalism and tourism industry transition to all-inclusive models.

Tourists are turning from an interest traditionally limited to the three Ss of sun, sand and sea to a new curiosity for the cuisine of the Caribbean. On the whole, however, the Pacific Islands do not yet see their cuisine as a tourist attraction (Chapter 6). Making these linkages work has taken four decades of changing tourist tastes and a desire for local foods with improved specialist production by island farmers, with many of the innovators being expatriates or return migrants (Part II; Momsen 1998). The demonstration effect of tourist food tastes has also increased demand for fresh foodstuffs from local populations.

In the countries of the global South, tourism grew with the arrival of jet aircraft, which made long haul tourism possible from the 1960s. These countries soon saw tourism as the *sine qua non* of development and, especially for small islands, it has become the driver of development (Scheyvens and Momsen 2008b). However, even here, local food and unspoiled beaches and coral reefs are major attractions. Economic relationships between tourist hotels and farmers are seen as a way of bringing together the colonial major economic sector of plantation agriculture with the new economy of tourism (Part II), but only if it can be done without harming the environment. Tourism industry consumption of water makes it less available to farmers for irrigation, whereas the use of agricultural chemicals pollutes the water supply, so sustainability of both sectors is becoming more difficult (ibid.).

In this book, several authors suggest the potential for enhancing synergies between rural and agritourism and environmental conservation. Rilla (Chapter 12) noted that agritourism entrepreneurs in California and England seemed to be better stewards of the environment and suggested that on-farm tourism income could indeed relieve some pressure on the environment associated with certain farming practices. Cox *et al.* (Chapter 10) noted that in Italy, the conversion to organic farming methods was associated with a parallel growth in agritourism and associated speciality food production targeting niche market consumers seeking high-quality, healthy, environmentally responsible and regionally produced foods. Similarly, in California and the United Kingdom, Rilla (Chapter 12) noted that several farms involved in agritourism shifted to organic farming methods and production of organic value-added speciality products.

Several authors posit a correlation between green, nature and sustainable tourism, and the demand for locally grown produce, resulting in enhanced opportunities for purchasing local food and for agritourism enterprises (Chapters 3, 5, 6, 10 and 12). Berno in her study in the Fiji Islands (Chapter 6) suggests that linkages between tourism and agriculture contribute to the ethos of sustainable tourism. She explains that offering 'sustainable cuisine', that is, distinctive regional cuisine based on locally produced food, could be an integral part of the sustainable tourism product and promotion. Also related is the growth in consumer demand for healthy food at a global scale and its impact on opportunities for agritourism initiatives and stimulating tourism industry demand for local food production (Chapters 3, 6 and 10). There is clearly a need for further research on the relationship between agritourism, demand for locally produced foods and the growing sustainable tourism market.

Farm stays are the focus of Chapters 5, 10, 11, 12 and 14 looking at Japan, Italy, California, New England, England and Spain. They usually involve extra work by farmwomen and are often seen as providing outlets for farm products especially for marketing local wine in California. They have also been thought to be a source of company for women living on isolated farms especially in Iberia (Canoves and Villarino Pérez 2002). Rilla notes that farm stays help to encourage younger members of the farm family to stay on the land by providing additional income and activities (Chapter 12). This is also seen in Hungary, where rural tourism is often a family activity (Chapter 2). Hashimoto and Telfer, writing on Japan in Chapter 5, look at the extra work running a bed and breakfast or farm-stays generates for farm women, where other members of the family rarely help. In Hungary, there are few farm stays because of the disconnection between land ownership and tourism following the post-Communist land tenure changes. However, a few farm stays are being set up in eastern Hungary, where rural tourism is less developed than in the west of the country (Chapter 2). In the global South, farm stays are rare because of differences in living standards between farmers and tourists. However, in Barbados, the Ministry of Tourism is encouraging the provision of private home-based accommodation for tourists in rural areas (Thangaraj 2009).

Farmers' markets, food festivals and the marketing of specialist products are seen as part of rural tourism in Hungary (Chapter 2), Japan (Chapter 5), Italy (Chapter 10), California (Chapters 11 and 12) and Barbados (Chapter 9). Wine tourism is of specific importance in California, Michigan, Italy and Spain and becoming significant in Hungary. Local indigenous brands of pork and cheese are beginning to be sold to tourists in Hungary (Chapter 2), whereas in Ghana, rural tourist sites are offering ostrich meat and the opportunity to make chocolate (Chapter 3). Production sites for these items have also become important tourist attractions such as the vineyards in California and Michigan, the rum distilleries in Barbados, the Oistins Fish Festival and the 'Taste of Barbados' culinary festival in Barbados, while sugar mills and banana museums are also major tourist attractions in many parts of the Caribbean. In Ghana, the site of the first cocoa plantation has become a tourist attraction (Chapter 3). Cox *et al.* (Chapter 10),

Rilla (Chapter 12), Donaldson and Momsen (Chapter 11) and Che and Wargenau (Chapter 13) in their case studies note that tourists, through such structured farm visits, gain exposure to new products which may, in turn, increase future demand for these products.

It appears that rural tourism in rich, industrialized countries and in poorer, developing nations is merging in character. In both circumstances, tourists come mainly from urban areas and increasingly visit the rural parts of their own countries or other nations. Rural tourism can be seen as a short-term reflux flow against the overwhelming movement of rural migrants to urban areas. Many rural tourists today may well be the urban migrants of previous generations looking for memories of their past rural idyll. Rural tourism represents a vehicle by which urban dwellers can experience the rural and satisfy a desire to connect to their food. Asiedu and Gbedema (Chapter 3) describe, in Ghana, how rural tourism presents an opportunity for children from cities to learn about farming and the origin of their food. Kukorelli (Chapter 2) argues that rural tourism in Hungary appeals to a desire by urban people to connect to their food.

The dearth of research on tourism and migration becomes even more striking with respect to the relationship between tourism, agriculture and migration. Evidence suggests, however, that the tourism industry, in particular urban tourist poles, can serve as a strong magnet for rural-to-urban labour migration, particularly when surrounded by impoverished and marginalized rural areas. Torres, in Chapter 4, demonstrates how tourism-driven migration to Cancún in the Yucatán has affected regional agriculture, with many farmers abandoning their villages and others attempting to maintain crops back home, while working low-paid seasonal wage jobs in the city. Although tourism has stimulated the emergence of new livelihood strategies, agricultural production and productivity have declined as a function of labour competition between the two sectors and the absence of farmers during critical periods in the cropping cycle. Torres suggests that this is particularly problematic when the high season for labour between tourism and agriculture overlap (Chapters 4 and 5). In other instances, tourism is also shown to be a force for slowing down rural-to-urban migration – notably in the case of agritourism, rural tourism and sustainable tourism among others. Demand for food from tourists, new jobs created in rural resorts and in farm stays, in addition to diversification that increases the profitability and viability of farms, may slow down rural out migration, particularly of youth (Momsen 2003; Chapters 2, 10, 12 and 14). Thus, tourism and agriculture linkages may have a dynamic impact on migration, while changing the nature of farming and of tourist visits. By examining these impacts in many different parts of the world, this volume provides global evidence for these changing relationships.

The issue of gender has not been addressed to a great extent in the tourism and agriculture literature, despite the fact that many of the agritourism initiatives rely heavily on female labour inputs and entrepreneurship (as mentioned in the earlier discussion of farm stays). In the case studies presented by Rilla in England and California (Chaper 12), also by Donaldson and Momsen (Chapter 11), Kukorelli in Hungary (Chaper 2) and Hashimoto and Telfer in Japan (Chapter 5), women were the primary agritourism entrepreneurs. These agribusinesses permit women to multitask with conventional reproductive activities and generate income through a home-based activity. Rilla suggests that there has been a feminization of farming through tourism activities that have served to diversify farm income and contribute to the bottom-line viability of small family-based producers. Hashimoto and Telfer discuss how agritourism serves not only to empower women on one level but also to inflict a considerable work burden on women who already put in longer hours than their spouses on other productive and reproductive activities.

One of the most interesting dimensions that emerges from this collection of chapters is the implication for rural familial relationships. Many rural tourism and agritourism enterprises are family operations, and several of the authors in this volume note that the additional income provided by the tourism activities results in younger generations remaining on the farm to work the family business rather than leaving in search of outside employment (Chapters 2, 10, 12 and 14). From a social perspective, this has the effect of keeping rural families unified, passing farming knowledge and rural lifeways to younger generations (Chapters 10, 12 and 14) and providing social care both for children (by grandparents) and the elderly (by their adult children and grandchildren) (see Chapter 12). Rilla also suggests that tourism can contribute to rural revitalization and impede the depopulation that is now occurring in many rural spaces across the world.

Much of our discussion is focused on the manner in which creating linkages between local agricultural production and tourism consumption, of both food and rural place, can create opportunities for farmers and the agrarian sector as a whole. This collection of chapters also demonstrates that integrating local food and rural places into the tourism product represents an important opportunity for diversification, revitalization, improved quality and expansion for the tourism industry. In Spain, Canoves (Chapter 14) traces how rural tourism has grown to become an important diversification strategy for rejuvenating the stagnating 'sun and sand' tourism upon which the Spanish industry was once almost exclusively based. Timms (Chapter 7), Rhiney (Chapter 8) and Berno (Chapter 6) argue that creating linkages to local agriculture, food and cuisine has enhanced the tourist experience and desire to experience authenticity, arguing that gastronomy is an important expression of culture and heritage. Several authors, including Che and Wargenau (Chapter 13) in their study of wineries in Michigan and Cox *et al.* (Chapter 10) in Italy, note that the consumption of regional products creates a unique sense of place, which is later evoked in the consumption of these products.

References

Canoves, G. and Villarino Pérez, M. (2002) 'Rural tourism, gender and cultural conservation in Spain and Portugal', in M.B. Swain and J.H. Momsen (eds) *Gender/Tourism/Fun?*, New York: Cognizant Communication Corporation: 90–108.

Momsen, J.D. (1998) 'Caribbean tourism and agriculture: New linkages in the global era?', in T. Klak (ed.) *Globalization and Neoliberalism: The Caribbean Context*, Lanham, MD: Rowman and Littlefield: 115–34.

Momsen, J.H. (2003) 'Participatory development and indigenous communities in the Mexican Caribbean', in J. Pugh and R. Potter (eds) *Participatory and Communicative Planning in the Caribbean*, Aldershot: Ashgate: 155–72.

Scheyvens, R. and Momsen, J. (2008a) 'Tourism and poverty-reduction: Issues for small island states', *Tourism Geographies*, 10(1): 22–41.

Scheyvens, R. and Momsen, J. (2008b) 'Tourism in small islands: From vulnerability to strengths', *Journal of Sustainable Tourism*, 16(5): 491–510.

Thangaraj, P. (2009) 'Tourism sector doing well', *The Barbados Advocate*, 10, 7 and 21 October.

Torres, R.M. and Momsen, J. (2004) 'Challenges and potential for linking tourism and agriculture to achieve pro-poor tourism objectives', *Progress in Development Studies*, 4(4): 294–318.

Torres, R.M. and Momsen, J. (2005) 'Gringolandia: The construction of a new tourist space in Mexico', *Annals of the Association of American Geographers*, 95(2): 314–35.

Part I

Tourism, agriculture and rural restructuring

2 Tourism and agriculture in Hungary: post-productivist transition or new functions in rural space?

Irén Szörényiné Kukorelli

Introduction

In Hungary, the communist regime collapsed 20 years ago, and there have been big changes in rural areas, especially in relation to land tenure. New activities such as rural tourism began to develop in the 1990s, but because of regional differences, this varies in nature between the more developed western regions of the country and the poorer eastern regions. This chapter examines the role and development of rural tourism and its linkages with agriculture in this post-socialist country and emphasizes the regional differences in relation to uneven development. In the more developed areas, successful tourism businesses are found near to national parks and spas, whereas in the less developed regions, rural tourism depends on local government assistance and is often related to local culinary festivals. Case studies illustrate these differences.

Post-1989 changes in the agricultural sector

Looking at the past 20 years of political, social and economic transition in Hungary, rural areas have experienced possibly even more significant social and economic change than urban places. Indeed, the 'rural change' process in Eastern European countries was both rapid and deep following the transition. Land privatization and the ensuing change in ownership structure and liquidation or transformation of socialist cooperative farms represented the first step of this process. Land privatization was achieved through a compensation process enacted by legislation, Act No. XXV of 1991, commonly called as the First Act of Compensation. This legislation established the principle that property losses caused after 1949 should be compensated, whereby 5 million Hungarian forints (HUF) was set as the upper limit of remuneration. Compensation tickets functioned as securities, which could be used for the purchase of land in competitive biddings in land auctions (Mihályi 1998).

Before the change from the communist regime, the land and property of agricultural cooperatives could have involved three kinds of ownership (Lovászi 1999):

- lands owned by the members of the cooperative,
- state-owned lands that the cooperatives used free of charge for an unlimited period of time and
- lands owned by the cooperative according to Act No. IV of 1967.

The land compensation began in 1992 and most had been completed by 1995. However, the process is still continuing today because of unsatisfied claims.

As a consequence of the Act on Cooperatives and the Act on Compensation and the subsequent compensation process, land ownership changed and the structure of Hungarian agriculture was transformed. The factors leading to the disintegration of the cooperatives were compensation on one hand and the distribution of the property among the individual members on the other hand.

Act No. II of 1992 led to the distribution of the property from 1,300 cooperatives, worth a total of 26 billion HUF (Csite and Kovách 1995). One third of the cooperative land was obtained by active workers employed in agriculture in the cooperatives, but whose families had owned land before collectivization, and 30 to 40 per cent by the pensioners of the agricultural cooperatives. The remainder was distributed to individuals who no longer farmed but whose families had owned land in the area before collectivization. Thus, cooperative land ownership ceased to exist. By 1997, 8 million hectares of arable land had shifted to private ownership, and landholdings became fragmented. Following the almost complete privatization of land, there are now nearly 2 million landowners in Hungary, and the average farm size was 3 hectares in 1999, according to Organisation for Economic Co-operation and Development data. A total of 11 per cent of all landowners had holdings smaller than 1 hectare, and 60 per cent of them had farms less than 10 hectares. This shows that the ownership and the cultivation of land are separate issues, and the conflict of interests of the two groups (landowners and growers) is still alive. This negatively affects the agrarian sector's competitiveness, impacts employment and income generation and has ramifications for rural society and spaces.

After the transition, the agriculture sector within the national economy lost its importance considerably; in 1989, its proportion of GDP was 13.7 per cent, but by 2005, it was only 3.7 per cent. In less than 10 years, production decreased by 35 per cent and stock rearing specifically by 50 per cent. In 1989, agricultural production contributed 22.8 per cent of the value of exports, but this figure dropped to only 7.2 per cent in 2005. The proportion of employment in agriculture was 22.8 per cent in 1989, but by 2005, it had fallen to only 5 per cent (Kiss 2002).

Such rapid and significant structural changes in agriculture accelerated the transformation of rural areas and emergence of new functions. Ilbery (1999) argues that the post-productivist transition is the next level of development and is characterized by the following:

- agriculture playing a decreasing role in employment and food production,
- importance of pluriactivity increasing,
- emphasis on quality food production increasing,

- continuous reforestation of arable lands and changing role of forestry in land use,
- penetration of small and medium enterprises, with the increasing role of high-tech and service sectors in employment,
- new uses of rural spaces becoming typical in the areas of small retail, tourism, recreation and protection of environmental values,
- repopulation of certain rural areas and
- increasing disparities in quality of life among rural people.

Many of the characteristics listed above are typical of Hungarian agriculture: the decreasing role of agriculture, the increasing proportion of forest, new emerging uses of rural space and increasing social differentiation and disparities in quality of life among rural dwellers. The question is whether it can be said Hungary has reached a period of post-productivist transition. The typical features can be seen, but the reasons for them are different from those in western European countries.

In the richer countries of the European Union, following reforms of the Common Agricultural Policy (CAP), protection of the natural environment and diversification have been encouraged through subsidies. Meanwhile, in Hungary, only domestic agricultural subsidies to support growers and production existed until 2004. However, the scale of domestic subsidies has been reduced, as indicated by the Producer Support Estimate (PSE)[1] value, which decreased by 10 per cent between 1996 and 1998 (during 1986–1989, it was 47 per cent), whereas at that time, this figure was 39 per cent (Kiss 2002) in European Union. Therefore, although the agricultural sector has been undergoing structural and land tenure changes, which have led to declining production and significant reduction in employment, the scale of state support has significantly decreased. As a result, incentives were missing which could have been directed towards the post-productivist agriculture preferred by the European Union. So, the signs of post-productivism, as noted by Ilbery, have been caused rather by the crisis of the sector and forced changes than generated by the actions of the CAP.

Rural development policy and its influences on rural tourism

It should be noted that certain incentives have resulted from EU actions in the form of the Poland and Hungary Assistance for the Reconstruction of the Economy, which encouraged community-based integrated rural development, including the establishment of rural tourism. In Hungary, the first form of community-based integrated rural development was the microregional associations initiated in the 1990s. This organization meant that local village governments or a small town with its surrounding villages established cooperation for spatial development. The population of these organizations would typically be approximately 10,000–15,000 inhabitants.

Tourism as a development objective was put forth in the first development concepts of microregional cooperation, and it was implemented by the establishment of accommodation sites and tourism products. Even before Hungary's EU accession in 2004, the Hungarian Ministry of Agriculture and Rural Development had subsidy programmes for agritourism; namely Vidék Fejlesztési Célelőirányzat (Target Allowance for Rural Development), which is a national subsidy financed by the Agriculture and Rural Development Ministry between 2000–2002. Further support has been provided from the Special Action Programme for Pre-Accession Aid for Agricultural and Rural Development (SAPARD) programme between 2000 and 2004. Indeed, 15 per cent of planned SAPARD support (€ 354,158) has been devoted to measure 'diversification activities and development economic activities ensuring alternative generative income'; however, the actual amount was only a thin slice of a total € 1,230,980. Both programmes are targeted at various priorities, including measure 1 promoting economic activities for alternative income earning (diversification) and measure 2 the revival and redevelopment of villages, and the protection and conservation of rural heritage, with the goal of stimulating the development of rural tourism in the civil, private and public sectors.

After Hungary's EU accession (May 2004), the National Development Plan was created for the period 2004–2007 involving five operational programmes. One of them was the Agricultural and Rural Development Operational Programme and the other was the Regional Operational Programme. Both of them set priorities for developing rural tourism. For the structural policy of the European Union, a new programming period started in 2007. For this new period (2007–2013), Hungary as all member countries has drawn up a National Development Plan, which involves the 'NEW HUNGARY Rural Development plan'. It must harmonize with the 1698/EC 2005 Regulation.[2] According to this regulation, Hungarian development policy directives, the so-called 3rd axis, set up priorities such as increasing rural income earning alternatives, and the 4th axis, that is, horizontal axis, is the LEADER Programme whose priorities are to develop local communities. The conservation of rural heritage and the promotion of rural tourism are listed among the activities eligible for direct assistance.

Despite access to the same forms of subsidization, there are marked differences between Hungary's different regions with respect to rural tourism. These differences have emerged for several reasons. Although the country's territory is small (93,000 square kilometres), several factors such as terrain, settlement structure, natural resources, forms of land ownership, social conditions and economic development generate significant differences. For example, these factors may explain why tourism in West Transdanubia has different specific features from that of the Hungarian Great Plain. Throughout the remainder of this chapter, I will present concrete examples to illustrate these differences in tourism development.

The differences between the rural tourism of the EU-15 states and Hungary are significant, given Hungary's later initiation of development and subsidy programmes. CAP funding is available for Hungary and the other new member states, but some of its regulations are not valid for the new member states. Still, the AGENDA 2000 Action Programme introduced the term 'modulation', meaning the opportunity to transfer direct payments made to farmers to be spent for diversification or other integrated rural development measures (Halmai 1995). This kind of modulation is not applicable for the new member states, thus Hungarian farmers and farm holders can use only earmarked subsidies designated for predefined purposes for the above-mentioned rural tourism development. However, direct payments paid to farmers, although still below the level of EU-15 members, are far higher than the amounts paid before the accession, and Hungarian farmers – unlike their West European colleagues – are not forced to look for alternative income resources to compensate for their lower income due to CAP reform.

Improvement and spatial differences of rural tourism

Regional differences in Hungarian rural tourism can mostly be explained by land ownership and settlement structure. In regions that are characterized by large landed property, the landowners/users neglect to develop diversification such as rural tourism. The Hungarian Great Plain is characterized by large property and by a farmstead settlement pattern where farm owners are not only landowners but also live primarily from agricultural production. Household stock breeding is typical with the reproduction of autochthon animal breeds for demonstration and gastronomic purposes. Unfortunately, a great number of farms are in poor condition and unsuitable for tourists because of the derelict state of buildings and their ageing proprietors.

Some so-called hobby farms also exist with a cottage, farm buildings and only a small piece of surrounding land, and they are used mainly as weekend or second homes. Hobby farmers sometimes provide lodging services for tourists in their renovated cottages. These areas, apart from a peaceful and quiet atmosphere, do not provide any significant tourist attractions. The neighbouring villages, small towns and agricultural communities are not tourist spots with the exception of settlements situated in proximity to Kiskunság and Hortobágy National Parks (Fehér and Kóródi 2008).

Several forms of rural tourism have developed in the western and north-western parts of Hungary with its historic villages, diverse natural landscapes and tourist attractions. This region is characterized by farmers with smaller parcels of land and citizens without land. Both groups deal with rural tourism. If we add to this the case of the West Transdanubian entrepreneurs who are engaged in the tourist industry, many of whom obtained their business experience in Austria, it is not surprising that rural tourism is more advanced in that part of the country (Table 2.1).

Rural tourism support schemes targeted civil organizations, local governments and even the private sector to meet the requirements of community-based integrated rural development. This was when the majority of tourism-related associations were established such as the Associations for Rural Tourism on a county level and tourist societies on a microregional level, such as the Society for the Tourism of Szigetköz.[3] These entities were founded to articulate local

Table 2.1 The relationship between rural tourism activities and landowners in Hungary

Entrepreneurs in tourist industry	*Geographical location*	*Activities*
Large farm holders	Great Plain	Farm tourism, breeding indigenous animal species, demonstration farming and gastronomical tourism
Small farm holders or landless entrepreneurs	West Transdanubia	Lodging
	North Hungary	Tourist services: ecotourism, spa tourism, equestrian tourism, biking tourism, canoe tourism and so on

entrepreneurs with members' microregion-related common market policy and also to submit grant applications for civil organizations to establish a regional brand organization and to fund training programmes. The public sector selected a different way for acquiring funding resources. Local governments created microregional-level associations and established their common development concepts and strategies. In addition to establishing rural tourism initiatives, they investigated all the components necessary for implementation, including increasing local identity, reviving folk traditions, identifying the place specific features of the local economy, preserving the location-specific traditions of handicraft industries and even the recovery of local gastronomy. Local governments and municipal associations acted as catalysts by assisting local societies and entrepreneurs in organizing special local events such as festivals, annual days of gastronomy or by building bike roads to give a further impulse to bicycle tourism. The tour paths built by local municipalities for ecotourism are also an example of the important role that local governments have played. Municipalities or municipal associations have also served as catalysts for the development of national parks, nature parks and even spas in the West Transdanubia region (Figure 2.1). In Hungary, apart from the central region with the capital Budapest, the West Transdanubian region is the second most developed region (10.7 per cent of the GDP is produced here) and has the second highest per capita GDP. The proportion of agricultural population is lower than the national average (4.2 per cent), and therefore, not surprisingly, the export of agricultural products from this region is also below the national average. This region is rich in varied natural attractions, due the fact that three national parks and five nature protection areas were established here over the past 20 years. In the Austrian cross-border part of the West Transdanubian region, the protected areas are linked to each other. In this region, from north-west to south, lies the Fertő-Hanság National Park established in 1994, the Sopron Nature Park with Sopron centre, the Írottkő Nature Park, then further to south along the border the Őrségi National Park established in 2002 and ending with the Hármashatár Nature Park up to the Slovenian border (Figure 2.1).

Figure 2.1 Rural tourism in study areas in Hungary: (1) Fertő-Hanság National Park, (2) Őrség National Park, (3) Kiskunság National Park, (4) Sopron Nature Park, (5) Írottkő Nature Park, (6) Őrség Nature Park, and (7) Hármashatár Nature Park.

All over the world, national parks are becoming targets for tourism, which is why the communities living on the periphery of these parks could potentially leverage their position to generate income (Ghimire and Pimbert 1997). However, national parks are sources of conflicts, as enforcement of rules and regulations by park authorities can violate the interests of the local population and of tourism and other enterprises. These two apparently contradictory interests should be reconciled. On one hand, the environment should be protected, preserved and improved, whereas on the other hand, the income and quality of life of the local population have to be increased through the development of tourism.

The partial results of cross-border research confirm that emergence of new functions are typical of a region's rural areas such as tourism, recreation, protection of environmental values and increases in the service sector (Momsen *et al.* 2005). The tourism product offered by the West Transdanubian region has been transformed in the past 15 years, affecting primarily the rural areas. National parks and nature park associations have been established by validation of a bottom-up policy. Certain regions have made or are currently making efforts to create rural tourism and different ecotourism activities through their local development policy.

Entrepreneurs engaged in the tourist industry should undertake the task of developing their own business, delivering investment projects, such as creating and expanding rooms for lodging, and accommodation and also providing special tourist services such as building equestrian riding grounds, creating a fleet of bikes, canoes and so on. The above-described trilateral partnership scheme may give a successful start to the development of tourism; thus, local governments have a catalyst role, and the civil sector plays a coordinating role in selling tourism as a product by organizing thematic (wine) tours and festivals, and they are also responsible for training programmes and common marketing campaigns. The role of private entrepreneurs here is accommodating guests and providing high-quality services for them.

The weakness of any of these three components will halt the tourism development process. If local governments are able to perform the role of catalysts but they lack the support of a strong civil organization to sell tourism as a product, then local entrepreneurs will be left alone with insufficient capital resources to supply an adequate product to attract tourists. Tourism will not mature if it lacks the support of the local entrepreneurial or business sector and thus has insufficient capital to implement development projects. This will have a negative impact on the tourist associations and societies as well. In such a case, although a 1-day tourist festival and programmes may be organized by a local community of villages, for example, due to inappropriate or absent accommodation facilities and services, they will remain only 1-day tourist attractions.

In non-protected natural areas in the region with spa and thermal baths, health tourism has started to develop because of the richness in thermal and spa water. In communities with these resources, as well as in neighbouring settlements, the development of services and hospitality linked to the spas and baths transformed traditional rural livelihoods. At first, settlements with thermal baths developed rural accommodations, but where one settlement could not fulfil the demand, more accommodations were established in direct neighbouring settlements. This happened in the cases of Hévíz, Bük and Zalakaros. Following the supply of accommodation, the traditional rural activities in this community have been changed to primarily service provision and hospitality. Some of the typical characteristics of protected areas and health tourism include

- leisure and relaxation in nature,
- strengthening environmental conservation,
- development of the service sector and of new professions,
- protection and education regarding heritage through marketing approaches and
- suburbanization and nascent enterprises connected to rural areas.

In these small rural areas (microregions), new rural tourist activity has become an important source of additional income since the 1990s. Rural tourism appeared as a development possibility at the national and regional scale in several planning and rural development documents in the first half of the 1990s. The rural tourism activity already operating before 1990 was based primarily on traditional leisure areas such as Lake Balaton. Since then, a new supply has emerged and continues to improve with small settlements offering the rural idyll, the nature protected areas offering natural heritage and intact nature and the settlements positioned directly on or neighbouring national or nature parks offering ancillary services.

This latter group includes the conditions of creating tourism, but the infrastructure conditions were incomplete or simply missing. Development of the infrastructure has enabled rural tourism in these microregions to increase the number of guests and consequently, both the number of tourists and overnight stays are rising. Those who offer programmes, in addition to accommodations, stand to benefit the most. It is a significant difference that although in Western Europe, and in Poland, rural tourism contributes to farm diversification, in Hungary, rural people (not necessarily those involved in agricultural activity) take on tourism activities as a way of diversifying rural economies (Momsen and Kukorelli 2007).

Leisure activities related to rural tourism include both traditional and new activities. Although the elements of traditional tourism were relaxation, passivity and low interest in getting familiar with the environment, modern rural tourism is active. Rural tourism today uses new marketing approaches and thus needs skilled professional personnel. It requires new activities, approaches, technologies and knowledge, not only from the demand side but also with respect to supply, which is accomplished through training and lifelong learning. Those kinds of enterprises appear to be generating new workplaces, expanding rural employment in ways that did not exist earlier.

The nature-based tourism offered by natural areas and national parks could be combined well with thermal and spa tourism, as this combination strengthens these two types of tourism while assisting the restructuring of rural areas. As mentioned above, after the collapse of cooperatives in Hungary, the released workforce was not able to work in the new activities. Only a small proportion of former agricultural employees, who were generally under qualified, became entrepreneurs. In most cases, they became unemployed or employees of new or restructured industrial firms and building enterprises.

The owners of rural accommodations today are well-educated people; in many cases, they were women working in town halls as administrators, or were teachers, or men with secondary or higher agricultural qualifications who have acquired enough information and capital to start an enterprise. Even people dealing with popular equestrian tourism have been middle- or high-level managers with qualifications in former agricultural cooperatives or young people who moved out from urban areas to live a 'new life' with the help of a new enterprise. In many cases, they wanted to realize the rural idyll (Murdoch *et al.* 2003). For example, in western Hungary, there are horse stables owned by a jeweller and real estate agent couple who moved from Budapest. Also, there is a similar horse stable, which was started by an agronomist, and today, his daughter runs the business.

Tours are supplied by the staff of national parks or the owners of accommodation, which involve ecotourism activities such as gathering mushrooms, archery, riding carriage and canoe tours, which are recommended and organized by the accommodation owners. In national parks, tourists come for bird-watching tours led by qualified national park staff.

Provision of new activities related to wellness and health tourism services has influenced the local workforce. Jobs such as hairdressers, cosmeticians, gift shop retailers, masseurs and dental assistants were not typical rural professions formerly. Today, these professions require continuous training and education.

Cases from the West Transdubian region

In the remainder of this chapter, I present an example of rural tourism development from the above-mentioned Szigetköz microregion in the West Transdanubian region (Figure 2.1). This area is located between the main channel and the side branch of the Danube River in north-western Hungary. This microregion has 26 member settlements (villages) in its territory. Local municipalities established a municipal association for lobbying and regional development in 1992, under the name of Szigetköz Municipal Association. This entity, engaged in the protection of environment and nature conservation, took the opportunity to develop grant applications to build a cycle track that could be part of the European bike track system lying along the Danube. The idea was to attract tourists to discover the value of their hidden natural resources. The association aspired to win a title of 'national park' or at least a nature park. The territories of Szigetköz Nature Conservation Areas on national level were delimited by the Ministry of Environment and then local municipalities built the eco-education paths in their territories. As a result of a joint venture, the bike track was built in several phases, which became a part of the Eurovelo cycle track system. The role of catalyst has been achieved. As far back as the 1990s, entrepreneurs in the tourism industry established the Association for the Tourism of Szigetköz. The association prefers ecotourism given the region's natural resources, and views canoeing, equestrian activities and cycling tourism (after the construction of the cycle track) as the principal areas of potential for tourism development. At the annual 'Travel Festival', organized nationwide in Hungary, and also famous in Europe, the association, as an independent body, publicizes the local tourism of Szigetköz.

Entrepreneurs in the tourism industry are not primarily landowners, but they would like to test their entrepreneurial skills by availing themselves of the opportunities presented by the change in regime. In several cases, it was an 'escape forward' from the threat of unemployment, and in other cases, venturing into tourism meant strong challenges to do something new, something different. The remainder of this chapter focuses on projects that seek 'a new way' of implementing tourism, rather than typical examples of diversification.

The following cases illustrate tourism businesses in Szigetköz, the north-west cross border of Hungary.

Katalin

Katalin was 40 years old when she built a guesthouse catering to tourists in the early 1990s. At that time, her husband had another job as an engineer in a factory in the nearby city of Győr and so was not involved in the tourist activities. He later lost his job and so he joined his wife's business, but his role is secondary to that of his wife. The guesthouse provides accommodation mainly for canoe and bicycle tourists because it is near to both the river and bike paths. Katalin is still active and a dominant member of the tourist association.

Marika

Marika was mayor of one of the villages and is a teacher. She promoted tourism activity and the marketing of the area not only because she was the mayor but also because she was president of the association of local governments in Szigetköz. She encouraged tourism in the village and surrounding area and provided accommodations for visitors. Her husband participated as a guide for canoe tourism, whereas their son ran a rest stop and small leisure centre along the bicycle trail. They own 6 hectares of land and this business is not their main source of income.

Ági

Ági is an energetic entrepreneur. Before 1989, she was a tax officer in the local government, but when the opportunity arose, she wanted to work for herself. Now she owns apartments that she rents to tourists. Her main interest is in bicycle tourism and she rents out bicycles to visitors. In addition to increasing accommodation capacity, she sees good perspectives in biking tourism and thus expanded her business with the assistance of resources she was granted from the SAPARD funds. Her husband helps out on weekends in his spare time, mainly with boating activities. He has been working in Austria for several years and provides reliable income for the family. As he works in the construction industry, he built the new apartments for visitors.

Márta

Márta was a nursery school teacher under socialism and her husband was manager of the local cooperative farm. After the transition, she lost her job and started a tourism business. Her husband became a large landowner through the compensation process and has his own businesses that are quite separate from tourism. In this household, rural tourism is not seen as being related to agriculture. In addition to ecotourist services, this inn provides gastronomic services for the guests by baking bread and roasting pork and other foods in its own oven. Although her husband has large landholdings, this catering service is provided on one of the streets of the village in the parents' renovated house, and no signs of farming can be seen. She is also an active member of the Association for the Tourism of Szigetköz.

The Szelle Horse Ranch

The Szelle Horse Ranch is a typical family-owned business with traditional gender roles. The husband looks after the horses and deals with visitors wanting to ride. His wife has a purely domestic role. Their daughter lives with her parents, she studies at the Pannon University in the agriculture faculty and helps in the business as a guide for the horse-riding tours. The qualified horse ranch and guesthouse welcomes guests from early spring to late autumn, prolonging the otherwise short tourist season by attracting visitors from Germany and Sweden. The family has expanded its business and now they lease canoes as well. In addition to running a tourist business, the family is a joint proprietor of a wood-processing company specializing in manufacturing modern insulated doors and windows.

Clearly, the roles of the various family members are dependent on the type of tourist facilities offered, and the relative employment needs and interests of both the wife and husband. Sometimes the business interests of family members may be separate and unrelated, whereas in other cases, they may be separate but related, and in another variation, they may all work together in the same business taking on different gendered tasks.

Cases from eastern and southern Hungary

In North Hungary, the region (Figure 2.1) with the lowest GDP per capita rate of all the seven regions of Hungary, local municipalities have sold tourism as a product, but entrepreneurs are the weakest links in the development chain. It is necessary to improve the quality of tourism investments; however, enterprises have low capital resources, and in several cases, they are unable to fund their own contribution. For this reason, they have no access to available subsidies. Those who are in need of extra income from tourism have no capital resources, and large landowners are not willing to undertake the risks of the tourist business that would provide only a small extra income for them, so there is little incentive for them to do so.

In this case, the catalyst role of local municipalities cannot be performed because although tourism has been created, there are no local entrepreneurs who can provide high-quality accommodation facilities and services for visitors. In such cases, 1-day tourist attraction events and festivals are organized. These festivals are good for the mobilization of local village inhabitants; they even attract visitors, mainly domestic ones, by offering a chance to taste local cuisine but unfortunately nothing more. With rising salaries, it is possible that some businesses can build their activities into a tourism product.

There are two interesting cases from the east part of Hungary where the local government foments tourism events without strong tourism enterprises. The first is the Nagydobos Pumpkin Festival that is held annually on the mayor's initiative. Nagydobos is a village in the Northern Great Plain region (Figure 2.1). The festival is organized as a part of the Village Day, which attracts an increasing number of visitors, with about 10,000 on a regular basis. Local residents present their products prepared in groups using their own recipes. This tourist product could be linked with further services so that it could extend the length of visitor stays. In another example, the Chef for Kings Gastronomic festival was initiated by the local government in Greatchef. Currently, it has grown into a 2-day festival, but it is still coordinated by the local mayor who inaugurates the festival with the mayors' culinary contest.

Finally, the 'Rendek Farm', where the Rendek family resides, is a good example of farm diversification through farm tourism. In this example, the landowner provides exemplary agritourism services on his farm by selling his own products to visitors as well. In this case, he received support and assistance neither from the local government nor from any association, but he had only his family land and a mission to preserve and transmit the rural lifestyle.

The 'Rendek Farm' lies in the Kiskunsági National Park (Figure 2.1), its buildings were built at the end of the nineteenth century. The family started farming again after the transition on 20 hectares around the farm buildings. They grow crops and breed animals, but their real mission is to look after and present the lifestyle of a peasant, so a farm museum was established in the old building. They produce indigenous breeds of animals (mangalica pig and racka sheep), and herbs are cultivated in the garden. Their aim is to sustain biodiversity and to carry out ecological farming, that is, to operate an organic farm. They produce their own natural products such as mangalica pork sausages and herbal tea, sweet and hot pepper creams, which they sell out of an organic product shop in the nearest large town, Kecskemét. In the farm museum, there are old farm instruments, old agricultural machines and an oven that was used for baking breads and frying meats by peasant families. However, the oven is not only an exhibition piece but is actually used, as the farm wife still prepares fresh breads, roast mangalica sausages or even dill and cheese pies for the guests in this oven. According to the host, the accommodation is 'in the hayloft', that is, it does not exist yet, but it is already planned. Their approach is to present peasant everyday life through the museum and gastronomy. They have good relationships with tourism accommodations operating in the surrounding area. Many guests come from places such as the capital city Budapest, 80 km away, and the nearest big town, Kecskemét.

Conclusion

Rural tourism after the transition has been rebuilt in Hungary and has led to the restructuring of rural space and society. Despite the fact that it is mostly seasonal and operates as an additional income-generating activity, its role is increasing in certain areas with natural beauty and richness in spa water. Formerly for the microregional associations of municipalities, today for the LEADER groups as in other EU countries, rural tourism as a priority of their own development plan plays an important role in almost every LEADER group. Of course, planning alone is not enough, demand for rural tourism must be generated as well. After the 1990s, with the exception of the Balaton area, tourism has been rebuilt from the ground up. Although the local stakeholders expected primarily foreign guests, clearly, as is common in Europe, it depends on domestic guests. Rural tourism in Hungary draws on the interest and wallets of the urban middle class, visitors who want to spend time on activities such as riding bicycles, riding horses, swimming, walking or other sporting activities or just to enjoy the special local gastronomy. The rural areas are continuously preparing themselves to welcome the guests. New developments have emerged in the rural tourism sector all over the country.

Over the past 20 years, rural tourism development has strengthened through partnerships with the municipalities and civil society (i.e. associations, national parks and other civil organizations) and entrepreneurs. Although the public sector is the so-called catalyst, the civil sector is working to create and market local tourist products, and the private sector contributes to tourism through its investments and businesses. Where the private sector is stronger and more skilled, rural tourism is better implemented. In the West Transdanubian region, as the second most developed region in Hungary, there are many successful examples of rural tourism. However, in the Great Plain region (Alföld), the catalyst role is stronger, but the private sector is weak and the supply of the tourism product is poor. In contrast to old EU member countries, in Hungary, rural tourism is not linked primarily to land ownership, but rather it is related to the class of rural society that is skilled and innovative; and is considered an opportunity to establish businesses after the transition, at first through providing accommodation and later other services.

Notes

1 PSE: all types of support are aggregated and indicate the rate of market price support.
2 Regulation of the European Commission.
3 Szigetköz is the north-western cross-border area of Hungary confined by the Danube and its riverside that is rich in natural resources, the development of its tourism began in 1990s.

References

Csite, A. and Kovách, I. (1995) 'Poszt-szocialista átalakulás Közép- és Kelet-Európa rurális társadalmaiban' (The post-socialist transition of rural society in Central and Eastern Europe), *Szociológiai Szemle*, 2: 49–72.

Fehér, I., Kóródi, M. (2008) *A vidéki turizmus fejlesztése* (The Development of Rural Tourism), Budapest: Szaktudás Kiadóház Zrt.

Ghimire, B.K. and Pimbert, P.M. (eds) (1997) *Social Change and Conservation Environmental Politics and Impacts of National Parks and Protected Areas*, UK: Earthscan Publications Limited.

Halmai, P. (ed.) (1995) *Az Európai Unió agrárszerkezete* (Agrarian-Structure in European Union), Budapest: Mezőgazda Kiadó.

Ilbery, B. (ed.) (1999) *The Geography of Rural Change*, Essex: Longman.

Kiss, J. (2002) *A magyar mezőgazdaság világgazdasági mozgástere* (The Hungarian Agriculture in World Economy), Budapest: Akadémiai Kiadó.

Lovászi, C. (1999) *Termőföld-tulajdonváltás Magyarországon 1988–1999* (The Change of Agricultural-Land-Owners in Hungary 1988-1998) Számadás a Talentumról sorozat, Budapest: Állami Privatizációs és Vagyonkezelő Rt.

Mihályi, P. (1998) *A kárpótlás* (The Compensation Process) (Számadás a Talentumról sorozat) Állami Privatizációs és Vagyonkezelő Rt, Budapest: Kulturtrade Kiadó Kft.

Momsen, J.H., Sz. Kukorelli, I. and Timár, J. (2005) *Gender at the Border: Entrepreneurship in Rural Post-Socialist Hungary*, Aldershot: Ashgate.
Momsen, J.H., Sz. Kukorelli, I. (2007) 'Gender and rural tourism in Western Hungary', *Eastern European Countyside*, 13: 83–96.
Murdoch, J., Lowe, P., Ward, N. and Marsden, T. (2003) *The Differentiated Countryside. Routledge Studies in Human Geography*, London and New York: Routledge.

3 The nexus between agriculture and tourism in Ghana: a case of underexploited development potential

Alex B. Asiedu and Tometi K. Gbedema

Introduction

Globally, tourism has grown to become one of the largest industries, generating foreign exchange and more direct and indirect employment than most of the traditional industries. According to the World Tourism Organization (UNWTO 2008), international tourism receipts rose by 1.7 per cent in real terms in 2008 to US$ 944 billion (€ 642 billion/£ 639 billion). This represents around 30 per cent of the worldwide volume of service exports and six per cent of overall exports of goods and services (UNWTO 2010). Tourism has also been viewed as a powerful tool for developing countries to trade their way out of poverty as these benefits are said to trickle down to the more peripheral regions, disadvantaged communities and the poor (Scheyvens 2007). In this regard, pro-poor tourism has emerged to emphasize the potential of tourism in achieving diverse development goals, especially within rural communities, which despite their rich endowment of prime tourist resources and attractions continue to lag behind in the development of national economies. Tourism may offer significant stimuli for many economic sectors and has significant backward linkages with major sectors like agriculture, industry and transportation (Torres and Momsen 2004).

As one of the more spatially diversified industries in the world, agriculture is able to integrate with tourism to generate increased growth and development in rural settings and poorer and marginalized sections of the society through pro-poor policy interventions. With the emergence of a new wave of rural and green tourism, there is a strong possibility that the position of agricultural or farm tourism may assume more prominence in consumer vacation decisions leading to the injection of a new source of ideas for tourism product development and marketing within farm-based tourism destinations.

According to Poon (1993), farm-based tourism, which has been used to describe agricultural or farm tourism, is a classic example of 'new tourism', even though Frater (1983) intimates that it is clearly not a new phenomenon. Further, Busby and Rendle (2000) have reckoned that tourism on farms becomes farm tourism as tourism revenue exceeds that for agriculture and farmers also start adopting business plans. According to them this transition is also realized as consumers begin regarding farm tourism as an enterprise. However,

Oppermann (1998) has suggested that this transformational process is often hindered by lack of experience of running tourism businesses, even though guidance for overcoming the challenges is available from various sources (see Chapter 14).

Ghana's tourism sector has expanded in terms of tourist arrivals and receipts over the past 20 or so years to become the third largest foreign exchange earner. Figures available from the Ghana Tourist Board (GTB) indicate that tourist arrivals increased from 125,162 in 1989 to 802,779 in 2009, representing an over five-fold increase (541 per cent) over this time period. Tourist receipts experienced an even larger increase over the period as they rose from US$ 72.1 million in 1989 to US$ 1,162 million in 2009. This represents a growth of over 16-fold (1,612 per cent) (GTB 2010). This expansion has been facilitated largely through increased promotion of the country's historical, cultural and natural resource assets that are predominantly rural based. The country's major attractions, in the form of game parks, rain forests, beaches, colonial forts and castles, rich cultures and others have been developed to propel the sector to become the third largest foreign exchange earner for the national economy. Political and economic stability has also contributed significantly to the achievements observed in the sector in recent years. Against this background, however, not much effort has been made at the level of policy formulation and implementation to integrate the tourism and agriculture sectors. Effective harnessing of the development potential of the combined action of these vital sectors is likely to lead to the resolution of some of the economic challenges facing some geographical areas and economic sectors of the country.

The focus of this chapter is to explore the interface between tourism and agricultural development in Ghana with a view to providing a better insight into the operations of two farm sites that now host tourists/visitors. This demand side analysis focuses specifically on the socio-demographic and travel characteristics of visitors and their motivations and levels of satisfaction associated with these trips. It further outlines the potential and challenges as well as recommendations for maximizing the development impacts of farm tourism, which remains largely an unexploited tourism resource base in Ghana. The chapter is made up of six sections. Following the introductory section is a literature review and a discussion of the current status of farm-based tourism in Ghana. Section four is devoted to the description of the study sites and study methodology whereas the next section provides a discussion of field data analyzed. Finally, we provide recommendations that outline strategies for overcoming some of the challenges identified in the study.

Literature review

The linkages between agriculture and tourism vary widely from region to region. Market factors that influence the relationship between the sectors are consumer demand (tourist and local), producer supply (agriculturalists and craftspeople) and food service providers (hotels, informal markets, restaurants and resorts) (Marcotte 2003). According to some authors, the relationship between agriculture and tourism has not always been symbiotic. Tourism has been associated with an increase in demand for imported food products (Marcotte 2003) and a decrease in local farm products, endangering local community survival. Although competition for land, water and labour has created many non-synergetic relationships, demand for tourism service provision has created interest in many agricultural activities. Tourism scholars report that in many instances, tourism has not only out competed agriculture for limited labour, land and water resources (Belisle 1983; Cox *et al.* 1995; Momsen 1998), but it has also been acknowledged that tourism, at the same time, has positively influenced the production of some agricultural products in many places. Thus, it can be argued that creating responsible linkages between tourism and agriculture shows promise for developing nations, in particular, because a substantial portion of their production and communities is intimately linked to agriculture (UNWTO 2008).

As a region, Africa is underdeveloped as a tourism destination for international tourists (Weaver and Fennel 1997). This is especially so with sub-Saharan Africa, which has begun to see a slow growth in tourist arrivals (0.8 million in 2008) (UNWTO 2010), as the overcrowding of several traditional tourism destinations, as well as improvements within the industry in many African states have made the continent a new target destination for tourists. Konadu-Agyemang (2001: 190) ascertained that 'in spite of the fact that Ghana is endowed with significant numbers of cultural and physical attractions in the form of rainforests, game parks, beaches, castles, and the like, the country as a whole has failed in the development of accessibility and accommodation'. Significant tourism assets and resources are remotely located, but their rural location need not be a reason for limiting their importance in national growth and development. Tourism drives economic growth in ways that make it one of the best engines for job creation and development available to those developing countries that possess natural beauty and the relevant infrastructure but lack other significant exploitable natural resources.

The Ghanaian context

Agriculture remains the most important economic sector of Ghana even though tourism has become one of the leading sources of foreign exchange for the country. The agricultural sector employs more than half the population on a formal and informal basis and accounts for close to 40 per cent of the country's GDP (ISSER 2008).

The country produces a variety of crops in various climatic zones that range from dry savannah in the north to rainforest in the south and are cultivated in two broad farming systems: a traditional subsistence system and a cash crop system. The traditional system is dominated by food crop farming and may also include pastoral farming and fishing. It is organized along small scale and subsistence lines to provide food for family consumption and the occasional sale of the surplus on the open market to raise money for the purchase of needed items. Operators within this farming system employ a low level of technology (Obia

1997). Productivity is low, and high levels of poverty are prevalent among the farmers involved. Other challenges that confront the traditional system are poor rural infrastructure, land degradation, inappropriate technologies, inadequate research, poor pricing and marketing policies and a weak institutional framework. Crops usually grown under this system are cocoa, palm oil, rice, yams, plantain, cocoyam (*Xanthosoma* sp.), cassava, maize, banana, vegetables and fruits. The commercial or cash crop sector involves the cultivation of crops, fishing and livestock farming on a large scale for local and external markets. This sector is characterized by capital intensity and the deployment of modern methods of production involving the use of machinery, on-site storage and processing plants and the provision of on-site residential accommodations for workers. The two systems are not mutually exclusive as the same farmer can be involved in traditional as well as cash crop or commercial farming (Obia 1997).

Cocoa from Brazil was brought to the Portuguese West African islands of São Tomé and Principe in 1819 and these islands became major producers of cocoa by the end of the century (Walker 2009). It was introduced into Ghana in 1879 and soon became an important export crop. Forty kilograms were exported in 1891 rising to 40 million kg by 1911 (Gordon 2009). Today cocoa generates over a quarter of Ghana's exports and Gross Domestic Product. According to ISSER (2008) rising prices in the mining sector and a fall in earnings in the cocoa sector since 2003 has brought the mining sector (gold, bauxite and manganese) back as the leading export earner in Ghana, accounting for 43.3 per cent of the nation's export earnings. However, the cocoa sector continues to provide over a quarter (26.3 per cent in 2007) of Ghana's export revenue and employs over three million households. Ghana is recognized as producing some of the best quality cocoa beans used to manufacture chocolate in the world (Daily Graphic 2008). Over the years, due to problems of soil infertility and pest infestation, cocoa cultivation has shifted away from the interior region to other frontier areas in the Western and Brong Ahafo regions of the country.

Tourism, on the other hand, could reduce disparities between urban 'cores' and their rural 'peripheries' (Konadu-Agyemang 2001). Growth in tourist arrivals and receipts has been rapid although from a low base with the exception of 2005 when a significant drop was experienced in the level of international tourist arrivals primarily because of apprehensions over the outcome of the December 2004 general elections. As can be inferred from Table 3.1, cumulatively, the increase for the period 1995–2007 was 73.7 per cent, representing an average annual growth rate of about 5.7 per cent over this period. Most of these tourists were overseas-based Ghanaians (11.4 per cent) and visitors from various countries including United States (10.7 per cent), Nigeria (9.6 per cent) and the United Kingdom (6.3 per cent); and the main purpose of these visits was visiting friends and relatives (25 per cent), doing business (22 per cent) and holidaying (20 per cent). In terms of tourist receipts, the increase was even greater. Tourist receipts raised by 177 per cent over the 1996–2007 period, that is, an annual growth rate of 13.6 per cent. However, tourism in Ghana still faces a number of challenges: poor transportation networks, especially those in rural areas, lack of good quality and basic

Table 3.1 Growth in international tourist arrivals and receipts in Ghana, 1995–2009

Year	*Tourist Arrivals ('000)*		*Tourist Receipts (in millions, US$)*	
	Number	*% Change*	*Total*	*% Change*
1995	286.0	5.4	233.2	2.5
1996	304.9	6.6	248.8	6.7
1997	325.4	6.7	265.6	6.8
1998	348.0	6.9	284.0	6.9
1999	372.7	7.1	304.1	7.0
2000	399.0	7.1	386.0	26.9
2001	438.8	10.0	447.8	16.0
2002	482.6	10.0	519.6	16.0
2003	530.8	10.0	602.8	16.0
2004	583.8	10.0	649.4	7.7
2005	428.5	−36.0	836.1	28.7
2006	497.1	13.8	986.8	18.0
2007	586.6	15.3	1,172.0	18.8
2008	698.1	19.0	1,403.1	19.7
2009	802.8	15.0	1,615.2	15.1

Source: Ghana Tourist Board (2010).

tourism infrastructure, inadequate skilled labour, leakages of financial resources to overseas destinations for the payment of imported goods used by tourism facility operators and a weak domestic tourism sector (JICA 2005).

Like in many West African nations, festivals are important events that attract many citizens living at home and abroad. These festivities account for a large part of the culture and heritage of the people and have become symbols for a variety of tourism attractions in the nation. A number of these festivals are devoted to the celebration of achievements in farming activities. Among them is the Rice Festival, celebrated by the peoples of Akpafu and the Yam Festival, Dzawuwu, celebrated by the peoples of Dabala in the Volta region. This latter festival is an annual traditional thanksgiving festival for the people of Agave. Others, like the Ngmayen of the Krobo people, the Ekyen Kofie and Kuntum festivals of the people of Sekondi in the Western region and of Enyam-Maim in the Central region (Ghana Tourism Homepage 2010) also relate to harvesting of one or two agricultural products. Festivals in Ghana are also known to contribute to the realization of certain development objectives as both residents and non-residents organize fund-raising activities to raise monies towards the development of vital projects within communities (see Asiedu *et al.* 2009). Nevertheless, the connection between such festivities and tourism remains either weak or underdeveloped.

Due to the lack of requisite management skills, finance, investments and infrastructure, only a very limited number of farms and farming related activities have been developed for tourist purposes in Ghana. This may be attributed to the fact that tourism itself has only assumed prominence in the last two decades and also that policies and programmes have not been adequately tailored towards the realization of the benefits that normally result from the interface between tourism and agriculture. Overcoming these challenges and expanding the tourism product that is generally offered by the country are some of the surest ways to achieve the targets set in the current national tourism strategy, which envisages among other things the realization of one million arrivals and one billion dollars by 2010 (GOG 1996).

The study sites and research methodology

Case studies were undertaken at two sites: the Tetteh Quarshie cocoa farm site at Mampong Akuapem in the Akuapem North District of Ghana; and Pacific Farms, a peri-urban farm site located at Klagon within the Tema Municipal Assembly Area, which is adjacent to the national capital, Accra (Figures 3.1 and 3.2).

The Tetteh Quarshie cocoa farm site, which is also referred to as the Eco-museum of cocoa in Ghana, is the pioneer cocoa farm in Ghana. The farm is about 58 km away from the national capital, Accra. Tetteh Quarshie, who established the farm, was a Ghanaian who lived on the island of Fernando Po (now called Bioko and a province of Equatorial Guinea and situated just to the north of São Tomé) in the nineteenth century. He returned to Ghana in 1876 with some cocoa seeds, which he used to establish the farm at Mampong Akuapem in Ghana (then Gold Coast) in 1879. Stories surrounding the history of this crop's development in Ghana claimed that Tetteh Quarshie swallowed the seeds to hide and transport them to Ghana, as rules on the plantation in Fernando Po where he was working forbade any employee from taking any seeds out of the plantation without authorization. Thus, back home in Mampong, Tetteh Quarshie went into hiding in the bush (now the present farm site) to discharge the seeds. He secretly protected this site to allow the seeds to grow. Three of the initial cocoa seedlings he planted on the site can still be found there even though most of the farm was replanted with new seedlings in 1960 by the local people. Visits to the site, which was then managed by the local people, witnessed some improvements after the replanting, albeit minor ones.

Following a discussion that took place in 2003 between the traditional chief of the area, the Ministry of Tourism and officials of the Ghana Cocoa Board, it was decided that the farm site be developed into a major tourist attraction to help improve visitation and increase revenue generation for the community. The meeting agreed that a museum be established where the full history of Tetteh Quarshie and the site would be documented and presented, and a reception centre be set up.

The site that hosts the annual national cocoa festival receives an average of 200 tourists per month. New reception facilities have also been constructed and plans are afoot to set up a small local chocolate manufacturing centre at the site so that tourists can be given hands-on experience in cocoa processing and also provided with the opportunity to taste chocolate they manufactured themselves. This adds more fun to visits, thereby enhancing the attraction. Located close to the farm site are other notable attractions such as the National Centre for Research into Herbal

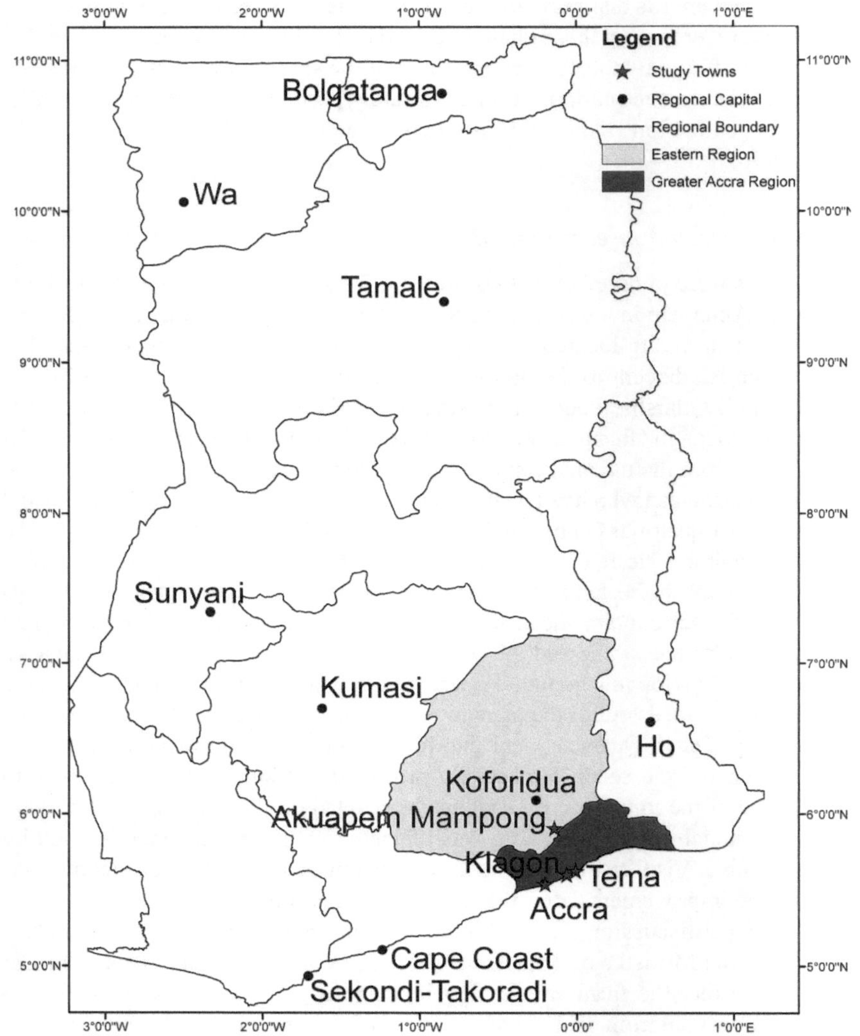

Figure 3.1 Map of Ghana showing the regions in which study farms are located.

Medicine, the Aburi Botanical Gardens and the Aburi Cultural Heritage site, where large quantities of tourist handicrafts and souvenirs are produced and sold. Other nearby tourist sites include some historical buildings associated with the introduction of the Presbyterian Church into Ghana and the site of the first teacher training college in Ghana at Akropong Akuapem (Figure 3.2). The Tetteh Quarshie Memorial Hospital, which serves as a major referral hospital

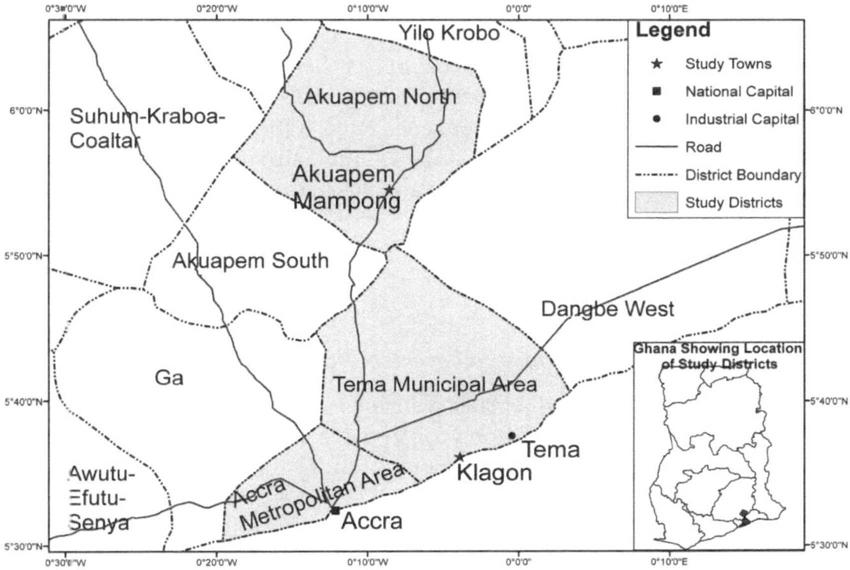

Figure 3.2 Map of the study districts and settlements in which farms are located.

in the area, is also nearby. The cocoa farm is presently jointly managed by both a private company and the Ministry of Tourism (Daily Graphic 2008).

The other site, Pacific Farm, is also an initiative of a Ghanaian expatriate but in this case he once lived in the United States. The farm was started elsewhere in the country (Afram Plains) as an ostrich farm and was relocated to its current site in 1995, a fast developing peri-urban neighbourhood located between the capital, Accra and Ghana's main industrial city of Tema. The farm has over the years expanded to include other poultry (fowls, turkey and ducks), pigs and fish farming. Sales of farm products such as meat, eggs, fish and feathers are conducted both on-farm and off-farm, especially at homes, hotels and restaurants. With the setting up of an on-site restaurant in 2000, visits to the site to have meals and view the ostriches have increased. Further development of the site was undertaken in 2003 with the construction of visitor accommodation facilities, especially for tourists visiting the metropolis and wanting out of town accommodation. Recent on-site developments have included the inauguration of a restaurant and a hotel. Regular visitors have included school children and students on excursions and families involved in weekend visits. Other visitors have included people from some of the neighbouring countries in the region who come mainly to buy products from the farm. In July and August especially, the farm also receives a small percentage of non-African visitors. On average, the site has about 300 visitors every week.

Field data were collected at the two sites in September–October 2008. The respondents represent a sample of visitors to the two sites who agreed to interviews using both open- and close-ended questions. Conventionally, this purposive, non-probability sampling approach has been found to be ideal for soliciting information from respondents visiting such facilities (Bryman 2001). In all, 56 questionnaire interviews were conducted at the two sites, 25 at the Tetteh Quarshie cocoa farm site and 31 at the Pacific Farm site. Questions posed in the questionnaire covered demographic variables, motivations for the visit, satisfaction levels and future plans.

Results

Socio-economic and demographic characteristics of visitors

An equal number of males and females were interviewed at the cocoa farm. More than 80 per cent (83.4 per cent) were young people aged less than 35 years. Representing a third of interviewees, the age range of 30–34 years was the dominant group (Table 3.2).

The majority of the visitors, in terms of educational attainment, were either university students or from other tertiary level institutions or students who had already completed their studies at that level (66.7 per cent). A further one-third (33.8 per cent) of the respondents were educated to Secondary/Junior Secondary School levels. Public service (25.0 per cent) and private, formal sector firms (25.0 per cent) constituted the main areas of employment of these visitors. Most of the visitors were, however, students (33.3 per cent). In terms of nationality, all of the respondents were Ghanaians with most of them residing within the Akuapem south district but outside Mampong Akuapem Township where the farm is located.

In contrast, the visitor sample of the Pacific Farms was dominated by males (61.3 per cent) who were aged 45 years and more (29.0 per cent). In terms of education level, data collected at this site showed an equal representation for the university and other tertiary level graduates (32.3 per cent) and for the Junior Secondary and Secondary School students (32.3 per cent). Additionally, visitors to this site were mostly self-employed people (32.3 per cent) and those engaged in the informal sector such as beauty shop operators, petty traders and car mechanics (25.8 per cent). Close to 94 per cent (93.6 per cent) were of Ghanaian nationality and resided mainly within the Tema Municipal Assembly area (77.4 per cent) but outside Klagon, where the farm is located.

Travel characteristics and motivation

Visitors to both sites relied mainly on information provided by friends and relatives (Table 3.3). At the cocoa farm site, a quarter of the respondents (25 per cent) claimed that their source of information about the site was friends and relatives whereas this was true for more than half (54.8 per cent) at the Pacific Farm site. The use of personal car (67.7 per cent) and public transport (32.3 per

Table 3.2 Socio-economic and demographic background of farm visitors

Characteristic	*Cocoa Farm*		*Pacific Farm*	
	Number	*Percentage Involved*	*Number*	*Percentage Involved*
(a) Sex				
Male	12	50.0	19	61.3
Female	12	50.0	12	38.7
Total	24	100.0	31	100.0
(b) Age (in years)				
15–19	6	25.0	5	16.1
20–24	4	16.7	2	6.5
25–29	2	8.3	8	25.8
30–34	8	33.3	4	12.9
35–39	2	8.3	1	3.2
40–45	2	8.3	2	6.5
45 and over	0	0.0	9	29.0
Total	24	100.0	31	100.0
(c) Educational Attainment				
Non-formal	0	0.0	1	3.2
Primary School	0	0.0	2	6.5
Secondary Sch./JSS	8	33.8	10	32.3
Senior Sec. Sch.	0	0.0	8	25.8
University and other tertiary level Schools	16	66.7	10	32.3
Total	24	100.0	31	100.0
(d) Job Type				
Students	8	33.3	2	6.5
Teachers	4	16.7	1	3.2
Public servants	6	25.0	6	19.4
Self-employed	0	0.0	10	32.3
Retirees	0	0.0	2	6.5
Others	6	25.0	8	25.8
Total	24	100.0	31	100.0
(e) Nationality				
Ghanaian	24	100.0	29	93.6
Non-Ghanaian	0	0.0	2	6.4
Total	24	100.0	31	100.0
(f) Place of Residence				
At farm settlement or township	2	8.3	2	6.5
Outside township but within district	10	41.7	24	77.4
Within the Eastern region but outside this district	2	8.3	0	0.0
Within Accra	6	25.0	2	6.5
Other areas in Ghana	4	16.7	1	3.2
Outside Ghana	0	0.0	2	6.5
Total	24	100	31	100.1

cent) dominated the mode of transport to the Pacific Farm site. At the cocoa farm site, most visitors took a taxi (41.7 per cent) to commute to the site, followed by the use of public transport (33.3 per cent). There was a wide disparity in travel party size between the two sites.

Visitors to the Pacific Farm site came mostly in groups of one to five persons (96.8 per cent). Half the visits to the cocoa farm site were also in groups of this size (Table 3.2). There was also a sharp contrast in terms of time spent at both sites. The Pacific Farm site was used only for brief visits, that is, for people staying for between 1 and 2 hours. Visitors to the cocoa farm site, on the other hand, stayed for longer periods, with most of them visiting for between 3 and 4 hours (41.7 per cent). A quarter of them stayed for much longer, that is, between 5 and 6 hours. Few of the visitors to both sites planned to visit other nearby attractions. Over half of them (54.8 per cent) responded that this was their first visit to the Pacific Farm site whereas only a third of the visitors to the cocoa farm site (33.3 percent) provided similar responses. This suggests high levels of return visits to the farms, in particular the latter which has existed longer. Study results suggest that return visits will only increase in the future, as the overwhelming majority of respondents at both sites indicated they would visit again (91.7 per cent and 80.6 per cent saying so for the cocoa farm and Pacific Farm sites, respectively). Some of the respondents who intended to re-visit the site in the near future elaborated on this as follows:

> 'I shall re-visit during the Christmas festivities to buy ostrich meat for my family'
>
> 'I will re-visit very soon, in the next few weeks, to complete my tour of the site'
>
> 'Today's visit has been a hurried one'
>
> 'I shall visit here with my family during Christmas'

As expected, visits to both sites were motivated by several factors. As indicated in Table 3.4, visits to the cocoa farm site were motivated by three factors. Most of the respondents singled out the atmosphere at the farm site during the cocoa day celebration (37.5 per cent) as their main or primary motivation for visiting. Others (33.3 per cent) cited seeing the farm and cocoa trees that are more than 130 years old and also the opportunity to learn more about the man who brought cocoa to Ghana, including his living environment.

Primary motivations for visiting the Pacific Farm were more diverse but interrelated. An overwhelming majority of visitors was motivated primarily by the sight of the ostriches and also to buy ostrich meat. As part of the growing trend towards eating healthy foods, ostrich meat has become quite popular with some Ghanaians and that may have fuelled the motivation to visit the site. Other motivations indicated were quite similar, including viewing both fish ponds and ostriches and buying of ostrich meat, feathers, sheep meat, duck meat and fish. Close to seven per cent (6.5 per cent) however cited the opportunity to have an overnight stay at the farm site hotel as their motivation (Table 3.5).

Table 3.3 Travel characteristics of respondents

Characteristic	*Cocoa Farm*		*Pacific Farm*	
	Number	*Percentage Involved*	*Number*	*Percentage Involved*
(a) Sources of Site Information				
Friends and relatives	6	25.0	17	54.8
From school textbooks/school teachers	8	33.3	8	25.8
Live close by site/same township	10	41.7	6	19.4
Total	24	100.0	31	100.0
(b) Mode of Transport to Site				
Own car/friend's car	4	16.7	21	67.7
Public transport	8	33.3	10	32.3
Taxi	10	41.7	0	0.0
On foot	2	8.3	0	0.0
Total	24	100.0	31	100.0
(c) Travel Party Size				
1–5 persons	12	50.0	30	96.8
6–10 persons	4	16.7	1	3.2
11–15 persons	2	8.3	0	0.0
16–20 persons	2	8.3	0	0.0
20+ person	4	16.7	0	0.0
Total	24	100.0	31	100.0
(d) Time Spent at Site (in hours)				
1–2 hours	6	25.0	31	100.0
3–4 hours	10	41.7	0	0.0
5–6 hours	6	25.0	0	0.0
7–8 hours	2	8.3	0	0.0
Total	24	100.0	31	100.0
(e) Are you visiting other attraction sites near the place?				
Yes	8	33.3	3	9.7
No	16	66.7	28	90.3
Total	24	100.0	31	100.0
(f) Number of previous visits				
None (first time)	8	33.3	17	54.8
Once before	4	16.7	3	9.7
Twice before	4	16.7	2	6.5
Thrice before	4	16.7	2	6.5
Over thrice before	4	16.7	7	22.6
Total	24	100.0	31	100.0
(g) Are you revisiting in the near future?				
Yes	22	91.7	25	80.6
No	2	8.3	0	0.0
Not sure	0	0.0	6	19.4
Total	24	100.0	31	100.0

Table 3.4 Primary motivations for visiting the Cocoa Farm site

	Motivation	*Percentage Involved*
1	To learn more about the man who brought cocoa to Ghana	29.3
2	To see the cocoa farm and cocoa trees that are more than 130 years old	33.3
3	To enjoy the cocoa day atmosphere at the farm site	37.5
Total		100.0

Table 3.5 Primary motivations for visiting the Pacific Farm site

	Motivation	*Percentage Involved*
1	To view and buy ostrich	64.5
2	To walk around the fish ponds and buy fish	9.7
3	To stay overnight at the hotel	6.5
4	To see ostriches and fish pond	9.7
5	To see ostriches and eat at the restaurant	3.2
6	To buy sheep and fish from here	3.2
7	To buy duck meat and fish	3.2
Total		100.0

Impressions about farm sites

On the general level of satisfaction with their visits, respondents gave very high rankings at both sites. As can be gleaned from Table 3.6, the Likert scale-based rankings were very high and favourable. At the cocoa farm site, very satisfied respondents accounted for 37.5 per cent of all responses whereas half of them (50 per cent) ranked their visits as satisfactory. Only 12.5 per cent responded that they were neither satisfied nor dissatisfied with the visit. No respondent claimed to be either dissatisfied or very dissatisfied. Reasons assigned for these high satisfaction levels included the lessons learnt on the background history of the farm, the farm itself, especially the aged cocoa trees there and Tetteh Quarshie's family house and the antiquated household and farm implements that are on display there. Others cited the serene farm environment as well as the fun associated with the national cocoa day celebration.

Responses from the Pacific Farm demonstrated much greater levels of satisfaction than those obtained from visitors to the cocoa farm. Over 58 per cent (58.1 per cent) indicated that they were very satisfied with the visit whereas almost 39 per cent (38.7 per cent) expressed the view that they had been satisfied with the visit. Reasons adduced in support of these high rankings included the sight of ostriches, ducks, fish ponds and the opportunity to buy meat within the farm site, eat at the restaurant and also stay overnight in the on-farm hotel if the need arose. Another expressed source of satisfaction was with the location of such an agricultural

Table 3.6 General level of satisfaction with visits to the farm sites

	Level of Satisfaction	*Cocoa Farm*		*Pacific Farm*	
		Number	*% involved*	*Number*	*% involved*
1	Very satisfied	9	37.5	18	58.1
2	Satisfied	12	50.0	12	38.7
3	Neither satisfied nor dissatisfied	3	12.5	1	3.2
4	Dissatisfied	0	0.0	0	0.0
5	Very dissatisfied	0	0.0	0	0.0
Total		24	100.0	31	100.0

facility within the urban setting which is increasingly becoming a rarity because of the high levels of urbanization that are being experienced within the locality. Rural agricultural farmlands are being increasingly acquired for the development of urban residential facilities. This facility therefore makes it possible to bring children to view animals and birds as well as nature closer to their homes.

A litany of on-site problems were mentioned by farm visitors. On the cocoa farm, opinions on this issue included the following:

> 'The site is small and not properly fenced and provided with requisite facilities and services for making visits more enjoyable'
>
> 'There is generally poor publicity and marketing of the site'
>
> 'Local involvement in the management of the site is limited even though it is a public facility'
>
> 'Virtually no benefits accrue to the local community in terms of revenue sharing of proceeds from the operations of the site and also jobs for the local people'

At the Pacific Farm site, most of the problems cited revolved around lack of basic on-site infrastructural facilities and services and the disinterest shown by community residents in acquiring skills from the farm site for self-employment in agriculture. The following comments provide an overview of the problems mentioned:

> 'Lack of play grounds and recreational facilities especially for children who visit here with their parents and relatives'
>
> 'The level of on-site facilities needs to be expanded. This includes improvements in the level of sanitation facilities and increases in the number of farm birds to enhance the visitor experience'
>
> 'Residents in the community appear not to take advantage of the presence of this farm to acquire skills in the running of farming businesses. The youth in particular, roam about in the community complaining about lack of jobs when this facility could turn their fortunes around'

Table 3.7 Suggestions for resolving farm site problems

Pacific Farm site	*Cocoa Farm site*
The farm operator/owner should develop recreational facilities for children to reduce their boredom when they accompany us here.	Site needs to be expanded and protected well. People in the community could be mobilized to assist in this.
More dams should be built to facilitate the expansion of the farm.	The place should be developed into a multi-recreational centre capable of attracting more visitors to help boost local economy and create jobs.
Improved sanitation and more hotel accommodation are needed.	Part of the money accruing from the site should be devoted to the development of the town.
Training programmes will have to be organized for the local people so that they can acquire skills to take up challenges in farming. This will help reduce the number of unemployed youth here.	The site should be widely publicized to attract more visitors.
	A hotel and a restaurant need to be provided to increase the attractiveness of the site to generate more revenue and jobs.
	Local people should be involved in the management of the site.

Suggested solutions

Respondents were also asked to propose solutions to the above listed problems. Suggested solutions for the Pacific Farm site focused primarily on infrastructure improvements and training programs for local people. Visitors to the cocoa farm mentioned infrastructure needs as well, but their suggestions concentrated more on the need to connect with local people and communities with respect to facility management, employment and local economic and community development. Table 3.7 presents some of the most common suggestions mentioned by informants at both sites.

Conclusion

This chapter has sought to explore visitors' experiences with two rural tourism enterprises in Ghana. The field data were gathered from two GTB-approved farm sites and findings that emanate from the study include the following.

First, it appears that operators of these two farm sites have responded very positively to demand by opening up these sites for visitation. Visitor demand for certain specific services, including a restaurant, hotel accommodation, a museum and sale of on-site products have helped towards the diversification of these farm businesses into rural tourism enterprises. The businesses appear to be proceeding well and judging from the high level of satisfaction expressed by visitors to the sites, it appears that the future is bright. However, more needs to be done to deal with some of the complaints identified above including the small areal extent of the sites and the inadequacy of site facilities and services. Most of the visits were by young people who were also relatively better educated and came largely from nearby locations. The above traits of the majority of the visitors can be capitalized upon to encourage more visits by these well educated and young people to these and other similar sites in Ghana in the near future. Even though the study did not delve into visitor expenditures, it can be deduced from the relatively short duration of the visits to these sites by visitors that the corresponding expenditures may also be low. This is likely to have serious implications for the profitability in the running of the sites and the multiplier effect within the local economy.

Second, few commonalities were found in the responses provided by visitors at the two sites on their travel characteristics and motivation. With the exception of travel party sizes, intention to visit nearby attractions, and plans to re-visit in the near future, the other responses varied among the two farm sites. As expected, motivations for these visits also differed between the sites. These differences are probably related to one site being mainly of historic interest whereas the other was more related to animals and gastronomy. Clearly therefore the sites need to broaden the scope and expand their range of product offerings so that they are able to attract more people to enjoy the wide range of products that is available and make their businesses more viable.

The third point regards general levels of satisfaction with the sites. Respondents expressed high levels of satisfaction with their visits to the sites even though visitors to the Pacific Farm demonstrated much higher levels of satisfaction than those visiting the Cocoa Farm. On-site problems, inadequate infrastructure and limited community involvement with either the management of the site or the acquisition of skills for the enhancement of livelihoods were commonly mentioned at both sites. Recommendations aimed at addressing these issues were also subsequently provided.

Finally, there appear to be only tenuous linkages between the farm-based tourism sites and the surrounding communities and their residents. Mutually beneficial exchanges in terms of skills training, information and flows of goods and services between parties were virtually absent, denying the adjoining communities the development opportunities that are normally associated with the presence of such tourist sites. Efforts are therefore needed to integrate the activities of residents within the adjoining communities into the planning, development and management of such attractions so that these communities share in the benefits that accrue from the attractions. It is one of the surest ways to ensure that such sites are managed along sustainable lines.

These findings led to the following conclusions and recommendations. First, there is an urgent need for the GTB to improve its information on the requirements needed for registration of farm-based tourism facilities in Ghana. It appears that a significant number of farms can operate as farm-based tourist destinations but are ignorant of the registration procedures involved and potential benefits that may be associated. To date only 15 such sites have registered to operate as farm-based tourism destinations with the GTB. This is in spite of increasing farm expansion, provision of modern on-site equipment and facilities and services that are likely to serve as springboards for the modernization of farm sites that have capacities to receive visitors for an enjoyable touring experience (pers. comm., Director of Planning and Business, GTB, 6 February 2009). The effective exploitation of this potential is likely to lead to additional income for farmers engaged in the promotion of these farms and help them reduce the risks that are normally associated with farming in Ghana. In addition, it is recommended that basic support in the form of logistics, training in farm management techniques and marketing and promotion of tourist products be organized for farmers who are interested in tourism. Second, existing and potential tourist operators should include these sites in the planning of their itineraries. This can assist in expanding and diversifying the mix of tourist products that are offered to visitors and attract foreign tourists.

On-site facilities need to be expanded and improved upon. There is a need for better sanitation facilities, provision of fencing and recreational grounds for children, increase in the stock of birds and animals, the enlargement of the site and the provision and expansion of hotel accommodation and restaurants. This is aimed at making visits more interesting, adventurous and stress relieving, which after all, is the essence of recreation. The provision of these services is likely to lead to prolonged stay at the site by visitors, including overnight stays and increased frequency of visits leading to more earnings and creation of additional jobs. Also, expansion can help cater to future potential increases in demand by people with interests in green, environmental, farm-based and rural and other related tourism segments.

Finally, improved local participation and involvement can result in community empowerment and thus poverty alleviation and development. Community involvement, especially in the running of a state-owned facility at the cocoa farm, can result in communities enjoying some of the benefits accruing from the operation of the site. These benefits may involve revenue sharing and job openings. Also local people could benefit from skills training in farm management, handicraft manufacturing and operation of small scale tourism outlets like food and other sales points for visitors. This is one of the surest ways of exploiting the nexus between the two vital economic sectors towards an effective and more sustained development of marginalized and peripheral communities and people; thereby reducing the levels of dependency of these communities on outsiders.

In conclusion, it is clear that although this study was case specific, employing data from two farm-based tourism sites in Ghana, the theme investigated is relevant to many areas in the developing world where agriculture is the pivot of rural and national development and tourism resources also abound. Lessons learnt here can therefore contribute towards the enhancement of rural livelihoods and poverty alleviation through a better utilization of the largely unexploited development potential existing between tourism and agriculture within the studied areas and beyond.

References

Asiedu, A.B., Labi, A.K. and Osei-Tutu, B. (2009) 'Promoting partnership with traditional authorities in the identification and development of cultural heritage: An Asanteman – World Bank initiative', *Africa Today*, Indiana University Press.

Belisle, F.J. (1983) 'Tourism and food production in the Caribbean', *Annals of Tourism Research*, 10: 497–513.

Bryman, A. (2001) *Social Research Methods*, Oxford: University Press.

Busby, G. and Rendle, S. (2000) 'The transition from tourism on farms to farm tourism', *Tourism Management*, 21(6): 635–42.

Cox, L.J., Fox, M. and Bowen, R.L. (1995) 'Does tourism destroy agriculture?', *Annals of Tourism Research*, 22: 210–13.

Daily Graphic (2008) 'Tetteh Quarshie cocoa farm attracts tourists', *Daily Graphic Newspaper*, September 11, p. 20.

Frater, J. (1983) 'Farm tourism in England: Planning, funding, promotion and some lessons from Europe', *Journal of Tourism Management*, September edition, 4(3): 167–79.

Ghana Tourism Homepage, (2010) Official Website of Ghana Tourism. Online. Available at: www.touringghana.com/festivals.asp (accessed 17 October 2009).

Ghana Tourist Board (GTB) (2010) '*Factsheet on Tourism Statistics in Ghana*', Accra, Ghana: Ghana Tourist Board.

Gordon, B.M. (2009) 'Chocolate in England from Introduction to Industrialization' in L.E. Grivetti and H.-Y. Shapiro (eds) *Chocolate: History, Culture and Heritage*, Hoboken, NJ: Wiley.

Government of Ghana (GOG) (1996) *National Tourism Development Plan for Ghana, 1996–2010*, Accra, Ghana: Ministry of Tourism.

Institute of Statistical, Social and Economic Research (ISSER) (2008) *The State of the Ghanaian Economy in 2007*, Accra, Ghana: University of Ghana.

Japan International Co-operation Agency (JICA) (2005) *Technical Cooperation for the Tourism Development Project Through Strengthening Public and Private Partnership in Ghana Document*, JICA Technical Cooperation Project with Ministry of Tourism and Modernization of the Capital City, Accra, Ghana: JICA Office.

Konadu-Agyemang, K. (2001) 'Structural adjustment programmes and the international tourism trade in Ghana, 1983–99: some socio-economic implications', *Tourism Geographies*, 3(2): 187–206.

Marcotte, J.W. (2003) *Agriculture and Tourism Linkages on Nuie Island: Emerging Alternatives for Small-Scale Producers*, unpublished Thesis, Masters of Science in International Agriculture Development, University of California, Davis.

Momsen, J. (1998) 'Time as a constraint on change in small-scale agriculture', In J. Hirst, J. Overton, B. Allen, and Y. Byron (eds) *Small-Scale Agriculture*, Canberra: Australian National University, pp. 1–10.

Obia, C.G. (1997) 'Agricultural development in sub-Saharan Africa', In S. Aryeetey-Attoh (ed.) *Geography of Sub-Saharan Africa*, Englewood Cliffs, NJ: Prentice Hall, pp. 286–324.

Oppermann, M. (1998) 'Farm tourism in New Zealand', In R. Butler, C.M. Hall, and J. Jenkins (eds) *Tourism and Recreation in Rural Areas*, Chichester: Wiley, pp. 225–33.

Poon, A. (1993) *Tourism, Technology and Comparative Strategies*, Wallingford, CT: C.A.B.

Scheyvens, R. (2007) 'Exploring the tourism–poverty nexus', In C.M. Hall (ed.) *Pro-poor Tourism: Who Benefits? Perspectives on Tourism and Poverty Reduction*, Clevedon: Channel View Publications, pp. 121–41.

Torres, R.M. and Momsen, J.H. (2004) 'Challenges and potential for linking tourism and agriculture to achieve PPT objectives', *Progress in Development Studies*, 4(4): 294–318.

Walker, T. (2009) 'Portuguese Cacao Cultivation in Brazil and West Africa, circa 1580–1912,' in L.E. Grivetti and H.Y. Shapiro (eds) *Chocolate: History, Culture and Heritage*, Hoboken, NJ: Wiley.

Weaver, D.B. and Fennel, D.A. (1997) 'Rural tourism in Canada: the Saskatchewan vacation farm operator as entrepreneur' In S.J. Page and D. Getz (eds) *The Business of Rural Tourism: International Perspectives*, London: International Thomson Business Press, pp. 77–91.

World Tourism Organization (UNWTO) (2008) *Yearbook of Tourism Statistics 2007*, Madrid: World Tourism Organization.

World Tourism Organization (UNWTO) (2010) Tourism Highlights 2009. Online. Available at: http://www.unwto.org/facts/eng/pdf/highlights/UNWTO_Highlights09_en_HR (accessed 23 March 2010).

4 Life between the two *milpas*: tourism, agriculture and migration in the Yucatán

Rebecca Maria Torres

The relationship between tourism and agriculture in the Yucatán Peninsula is a function of the proliferation of Planned Tourism Development (PTD) as an economic development strategy and the implementation of neoliberal agrarian reforms (Torres and Momsen 2005b). The latter has led to a pronounced restructuring of rural communities, as households and individual farmers have developed a variety of survival strategies in their struggle to negotiate their position within a constantly shifting local context, influenced by the larger global political economy. A common strategy since the inception of mass tourism in Cancún, during the early 1970s, has been rural-to-urban migration with rural people moving to tourist poles on both a temporary circular basis and permanently. Rural-to-urban migration to the tourist poles, and more recently to the United States, has had a profound impact not only on Yucatán's agriculture (Radel and Schmook 2008; Schmook and Radel 2008) but also on all dimensions of rural life, including economy, culture (Re Cruz 2003), gender and familial relations (Bever 2002; Castellanos 2003, 2007; Re Cruz 1998), consumption patterns, and community and individual subjectivities, among others.

This chapter draws on field research conducted in a Cancún shantytown, the Franja Ejidal[1] and rural communities throughout the State of Quintana Roo from 1996 to 1998 and in 2003 (and continued contact with one case study family) and examines the processes and impacts of tourism-driven rural-to-urban migration on Maya agricultural households. Specifically, the chapter will shed light on the experience of Cancún migrant *milpero* families as they seek to negotiate their position between urban tourist and rural agricultural spaces. *Milpa* farming is the shifting cultivation corn farming system still employed by most Quintana Roo farmers. I draw upon Re Cruz's (1996, 2003) conception of Cancún as a metaphorical 'other' *milpa* in the geographical imagination of migrants who retain their Maya identity as *milperos*, in part, through the conception of Cancún as a distant *milpa*. As with making *milpa*[2] in the forest, wage labour in Cancún yields critical productive resources for the survival of the Maya.

The first part of this chapter examines the broader patterns, trends and impacts of tourist-driven rural-to-urban migration to Cancún from the urban perspective, drawing on a household survey I conducted in the Cancún *Franja Ejidal* shantytown where many immigrant families live. A random sample of 333 households was drawn from a transect strip extending from the now 'gentrified' *franja* (fringe) of older settlement, which is part of the basic infrastructure-supplied urban core, extending outward to the edges of the shantytown where new immigrants inhabit tar paper shacks. This strip was defined to ensure that the sample would include an even distribution across the Franja's temporal spectrum ranging from well-established neighbourhoods of two storey cinder block homes on paved streets to random clusters of cardboard and tarpaper shacks of new immigrants on the edges of Franja. In particular, the survey sought (quantitative and qualitative) information about the characteristics of migrants, origins, livelihood patterns, present conditions and connections with the families, farms and communities left behind. Also, in this section, I present data from a household migration survey from three villages (two in Quintana Roo and one in Campeche). This latter survey was administered to 449 households in these communities (in addition to 20 in-depth, semi-structured qualitative interviews) and consisted of a brief set of questions to establish household migratory links with the tourist poles, and identify the extent of emerging patterns of migration to the United States.

The second part of the chapter focuses on an in-depth case study of a Cancún *milpero* and his family to understand, at ground level, how tourism-driven rural-to-urban migration reshapes rural life – notably agriculture. *Milpa* farming remains the predominant and most widespread form of agriculture practised in the state. Between November 1996 and January 1998, I followed the full *milpa* cycle of management, employment and circular migration of a Cancún *milpero*, Don[3] Mario, and his immediate family. The farmer, assisted by one of his daughters, maintained a structured daily log recording all activities (and their duration) that might in any fashion be pertinent to his *milpa* enterprise. The log also included complete procurement and expenditure records. In addition, I visited the *milpa* site and conducted interviews at regular intervals sufficient to enable a photographic and video record of all phases of the enterprise. This included visits scheduled to coincide with key events such as onset of clearing, plot burning, planting and harvesting. Regular visits were also made to neighbouring *milpa* sites maintained by full-time farmers to provide a frame of reference.

Both parts of the paper are informed by a series of farmer and *ejido comisariados* (*ejido* leaders) semi-structured, in-depth interviews conducted in rural Quintana Roo and interviews with rural immigrant families in Cancún. These interviews were not only part of a larger study on tourism and agriculture linkages (Torres 2003) and tourism-driven rural-to-urban migration but they also shed light on the impacts this has had on agricultural communities and, more specifically, on rural families (Carte *et al.* 2010; Torres and Momsen 2005a). In particular, I draw on interviews with the *comisariado* and other farmers in the *ejido* Nicolás Bravo, Don Mario's village of origin, and the setting for the *milpa* case study to provide the backdrop for the story of Don Mario's struggle to make a living between the two '*milpas*' of Cancún and his fields near Nicolás Bravo.

While acknowledging competition for labour between the two sectors, the tourism and agriculture literature tends to ignore the issue of tourism-driven migration, notably the tendency for large-scale PTD to trigger local and national migration to the tourist poles. While this chapter discusses tourism-driven migration, it is important to recognize that, in some cases, as noted in Chapters 2, 10, 12 and 14 of this volume, tourism, especially agritourism, can also deter rural-to-urban migration by diversifying and increasing farm incomes to entice family members, particularly youth, to remain in rural communities. This chapter contributes to a gap in both tourist studies and the tourism and agriculture literature by elucidating the complex relationship between tourism, agriculture and migration at regional, household and individual scales. Cancún, established in the early 1970s as the first mass tourism master planned resort, is the ideal laboratory for examining the impacts of mass tourism development on agriculture and rural communities and households through initial production-led circular rural labour migration to the urban poles usually becoming permanent settlement in time.

Rural-to-urban migration

The diverse Quintana Roo agricultural landscape has recently undergone complex processes of rural restructuring as a result of neoliberal, political and economic changes occurring at the global, national and local scales. Pronounced rural-to-urban migration from the Quintana Roo hinterlands to the urban tourist poles has been an outcome of national neoliberal agrarian policies, and along with those policies, migration has also been a source of profound rural restructuring in the region. PTD, an arguably neoliberal development strategy for the peninsula, resulted in uneven development, exacerbating disparities between rural and urban spaces (Torres and Momsen 2005b). Perhaps, the greatest effect of PTD on the Yucatán rural landscape was that it spurred an exodus of peasant farmers to the urban tourist poles in search of employment and improved quality of life. In addition to the tourism monoculture approach to regional development, the dramatic cuts in farm subsidies, agrarian credit and other forms of farmer support have all pushed farmers to the city on either a temporary or permanent basis as a household survival strategy.

In Cancún and other Riviera Maya tourist resorts – the metaphorical *milpa* – most of the jobs available to peasants are low paying, low-skilled employment in construction, domestic service and in hotels and restaurants. Most of the jobs are temporary in nature, fluctuating with the tourism season and offering little employment security. Urban-dwelling rural migrants typically inhabit the most impoverished, under-resourced shantytowns on the resort periphery. The newest areas of the Franja Ejidal in Cancún – typically devoid of urban infrastructure such as water, electricity, roads and sewerage – house the poorest new arrivals in a mélange of makeshift shacks and tar paper boxes. Municipal governments of the tourist poles experiencing the most rapid urbanization, notably Playa del Carmen and Cancún, cannot keep up with their exploding infrastructure needs.

Studies from the 1990s suggest that tourism-driven rural-to-urban migration is perhaps the most powerful factor reinforcing rural restructuring in Quintana Roo and the Yucatán Peninsula as a whole (Kintz 1998; Pi-Sunyer and Thomas 1997;

Table 4.1 State of origin of immigrants surveyed in the Cancún Franja Ejidal

State	*Frequency*	*Per cent*
Campeche	25	7.5
Chiapas	29	8.7
Chihuahua	1	0.3
Colima	1	0.3
Distrito Federal	5	1.5
Guanajuato	2	0.6
Guerrero	11	3.3
Mexico	2	0.6
Michoacan	2	0.6
Oaxaca	7	2.1
Puebla	3	0.9
Quintana Roo	25	7.5
San Luis Potosi	1	0.3
Tabasco	42	12.6
Tlaxcala	1	0.3
Veracruz	37	11.1
Yucatán	138	41.4
Zacatecas	1	0.3
Total	333	100.0

Source: Immigrant Survey, 1997, $N = 333$.

Radel and Schmook 2008; Re Cruz 1996, 1998; Schmook and Radel 2008; Tolen 1998). Given the rapid and informal nature of this migration, there are few accurate, comprehensive migration statistics. My research, which spanned the mid-1990s to 2003, indicates that the migration of Quintana Roo's rural inhabitants to the expanding tourist poles, although initially temporary in nature, came quickly to embody both permanent and circular patterns of migration. Similarly, migration to the United States, until recently non-existent in most rural Yucatán villages, has now become a common phenomenon (Fischer 2007).

According to my survey of 333 households in the Cancún Franja Ejidal during 1997, only 7 per cent of the residents were originally from Quintana Roo (see Table 4.1). Residents of the Franja are generally permanent migrants. Temporary workers seek rentals, live with family members or have lodging provided by employers (particularly with construction work). This low percentage of Quintana Roo immigrants reflects both the temporary pattern of migration prevalent in the region at the time and the relatively small Quintana Roo base population.

Migration has not only affected rural Quintana Roo, it has had a similarly large impact on the hinterlands of neighbouring Yucatán state. According to the Franja Ejidal Migration Survey, 41.4 per cent of all residents were originally from Yucatán state. Similarly, a COPLADE/FONATUR study of Cancún residents in 1985 found that 45.8 per cent of the population was originally from Yucatán state (Fraga 1992). This suggests that more of the Yucatán migration, in contrast to

Quintana Roo, was permanent during the 1980s and 1990s. Re Cruz (1996) reported in her study of the Yucatecan village Chan Kom that 44 per cent of the migration to Cancún was permanent. The high level of immigration from Yucatán state is a function of various factors, the most obvious being geographic proximity to Cancún. The high levels of immigration are also, most probably, because Yucatán *ejidos* are much smaller than their Quintana Roo counterparts and, therefore, subject to more extreme population pressure. In addition, the arid and rocky soil conditions characteristic of the north-western peninsula are also less suitable for agriculture than in southern Quintana Roo. The state of Yucatán, which has been fairly densely populated since pre-Hispanic times, has also suffered the effects of long-term overexploitation without proper soil conservation. More than any other state in Mexico, Yucatán has been affected by rural-to-urban migration of peasants to Quintana Roo tourist poles. Migrants from the Yucatán have long emigrated to the tourist poles in search of work, with this internal migration sometimes serving as a 'stepping stone' for international migration (Rodríguez *et al.* 2007).

There are significant differences in migration patterns across Quintana Roo communities depending on geographic location, economic base and ethnic composition, among other community factors. Kintz (1998) in Cobá found that it was common for work groups of men to leave the community to work on construction and archaeological excavations on a temporary basis, with permanent migration being less common. Dufresne (1995) observed in a Dzonotchel survey that 33 per cent of the respondents migrated to work in tourist poles on a temporary basis. Approximately 14 per cent of the respondents declared their migration to be permanent. Based on field observations and interviews with local officials and academic researchers, it would appear that permanent migration from Quintana Roo rural communities was significantly less common than temporary migration during the first three decades of Cancún's existence.

Over time, many communities have developed a complex mix of both circular and permanent migration patterns to tourist poles and, more recently, external migration to the United States. In 2003, I conducted a household migration survey in three rural communities (see Table 4.2). Although permanent migration to the tourist poles had been relatively low in contrast to circular migration in previous years, by 2003, permanent settlement had become more common. The most surprising result was the high level of migration to the United States. In the mid-1990s, for example, when I began research in Nicolás Bravo, migration to the United States was virtually unheard of. When I returned in 2003, however, I found that over a quarter of the households had one or more members having migrated to the United States (see Table 4.2), while other communities in the Yucatán have had a longer history of migration to the United States. Morocoy, a community that I surveyed in 2003, has been referred to as '*el pueblo sin hombres*', or 'the town without men', with over half its households indicating that they had one or more members in the United States.

Migration, whether permanent or circular, has a significant impact on the agricultural landscape. Permanent migration often translates into abandonment of

Table 4.2 Percentage of surveyed households with migration by destination and type, 2003

Village	*Households with permanent migration to tourist poles*	*Households with temporary migration to tourist poles*	*Households experiencing some form of migration to tourist poles*	*Households with migration to the United States*
Morocoy (*N* = 200)	19.5	10	29.5	53.5
Nicolás Bravo (*N* = 200)	15.5	11.5	27	25
Veinte de Noviembre (*N* = 49)	14.2	2.04	16.3	16.3

farming altogether, whereas many temporary migrants attempt to continue farming activities in their villages of origin (Dufresne 1995; Re Cruz 1996, 1998). In the Cancún Franja Ejidal Survey, where most inhabitants were permanent or long-term migrants, 27 per cent of all respondents had been involved in farming or fishing prior to settling in Cancún as was true for the same proportion of Yucatecan settlers. Of the Quintana Roo settlers, 36 per cent had some experience working in agriculture. These numbers suggest that the majority of permanent Cancún immigrants are from other urban areas. This is consistent with the notion that the more educated immigrants generally coming from urban areas are able to capture most mid-level year-round jobs, whereas rural people often end up with lower paying temporary work. Over a quarter of Franja Ejidal residents are, nevertheless, people who have abandoned primary sector activities. Forty-five per cent (151) of Franja residents work in tourism industry jobs or ancillary and related services. Clearly, the tourism industry has pulled a significant number of workers out of the primary agriculture sector into the tertiary tourism service sector.

In rural areas, a reduction in available labour has necessitated a restructuring of agriculture and household management of farming activities. Re Cruz (1998) suggests, for example, that in Chan Kom, women are taking on an increasingly important role as decision makers in farming, while compensating for their husbands' absence with family or paid labour. She also noted that average *milpa* sizes have decreased over time as a function of out-migration. Dufresne (1995) observed in Dzonotchel that *milperos* invest less time in their plots, plant smaller areas and produce lower yields. In their study of a southern Yucatán community, Radel and Schmook (2008) found that transnational migration became important in the 1990s reflecting a shift in traditional agricultural livelihoods and, in particular, deforestation. In their study area, migration was part of a combined strategy consisting of cultivation of chilies as a cash crop, which, however, was perceived as risky because of environmental factors and pests. As more men migrated, chilie production declined, although the community continued to invest in agriculture.

One of the greatest impacts of tourism-driven migration on rural restructuring is its effect on labour costs. Typically, in subsistence or small-scale production, farmers do not place a monetary value on their own labour inputs. With tourism development, the cost of agricultural labour has risen to prohibitively high levels as rural workers expect wages comparable with what is paid in the tourism poles. Given the low yields and poor profitability of agriculture in the region, few farmers can afford to pay for labour. Additionally, the opportunity cost for farmers to work in their fields is high, given the alternative tourism wage market. These high real and opportunity costs have rendered agriculture in the region unprofitable.

The effects of both circular and permanent tourism-driven migration on rural communities have been significant. Migrants reconstruct their ethnic, racial and political identities as they negotiate their positions in the social fields linking rural agricultural space and urban tourist space. Following their arrival in the city, rural immigrants undergo significant changes in language (replacing Maya with Spanish), dress and consumption. Arnaiz and Dachary (1992) suggest that many immigrants initially maintain strong ties to their village, often returning to participate in agricultural activities during their first 5–10 years in the city. Under this scenario, immigrants gradually become permanent residents in tourist poles as their linkages to the city strengthen and their ties to the village weaken.

Migration in the Yucatán Peninsula is a highly gendered process. Often the earliest immigrants are men, leaving behind a growing number of female-headed households (Kintz 1998; Re Cruz 1998). This is not only because of male outmigration but also the result of an increasing economic independence on the part of women. Kintz (1998) noted in Cobá that tourism has allowed rural women to become entrepreneurs engaging in household-based businesses such as craft production and food preparation to serve the nearby tourism markets. This has given women a certain degree of economic independence, leading in some cases to the freedom to exit an unhappy family situation. However, in some areas, women represent a significant proportion of migrants. In Chan Kom, Re Cruz (1996) noted that 33 per cent of migrants were women (20 per cent permanent and 13 per cent temporary), with an increasing number being single mothers because their husbands have paired with other women (often in the tourist pole) and left their first families to fend for themselves.

Labouring between the two *milpas*: Cancún *milpero* life and cropping cycles

The following case study presents an in-depth look at the experience of a Cancún *milpero*, Don Mario, and his family as they struggle to construct a livelihood between the two *milpas*, Cancún and their home village *ejido* Nicolás Bravo, while keeping the family intact. To frame the case study, I present an overview of the *ejido* context based on field visits and interviews with the ejido comisariado and other farmers (including Don Mario). Then, I trace the milpa cycle over a year as a temporal frame from which to knit together the story of Don Mario and his family's attempt to work the two *milpas* for survival. Only through a detailed discussion of crop management is it possible to understand the subtle disruptions, disconnects and challenges Cancún *milperos* face as they circulate between their two *milpas*. In this manner, we can untangle the multiple and subtle ways in which tourism-driven migration and the state agrarian policy affect agriculture and, most importantly, the rural families with one foot in both worlds.

This case study is based on a series of interviews and field visits with Don Mario in both Cancún and Nicolás Bravo. Field visits were conducted during all key tasks, including during the *tumba* (land clearing), the *quema* (burning the brush), planting, harvesting and a harvest of honey from Don Mario's brother's hives. In addition, Don Mario, with the assistance of his oldest daughter, Raquel, kept a daily *milpa* management diary, including tasks performed, workers, area covered, tools employed, hours worked, expenditures and cash outlays.

The setting: Ejido *Nicolás Bravo*

The *ejido* Nicolás Bravo, located 65-km east of Chetumal, on the border with Belize was established 55 years ago as a *chicle* (a natural gum) camp by immigrants from the state of Veracruz. Nicolás Bravo currently has 486 *ejidatarios* (town population of approximately 3,653 inhabitants, Instituto Nacional de Estadistica y Geografía 2005) with rights to 88,450 hectares of land. Of this land, 51,150 hectares is deemed arable land fit for cultivation, 35,800 hectares is forested and 2,500 hectares is pasture. Nicolás Bravo is one of the wealthiest *ejidos* in terms of natural resources, not only because of its large size but also because it contains some of the best farm land and most productive forests in the state of Quintana Roo.

The *ejido* has a mixed population with approximately 50 per cent being of Maya descent, 30 per cent of which are Maya immigrants from Yucatán and 20 per cent original Maya inhabitants of Quintana Roo. Another 30 per cent of the population are immigrants from Veracruz and approximately 20 per cent are originally from Tabasco. The community began as an itinerant *chiclero*[4] camp, with workers and their families gradually staying on and homesteading. The first influx of immigrants was made up of land-hungry peasants from Veracruz. In addition to the 486 *ejidatarios*, who hold titles to plots of land, there are numerous '*repobladores*' – people who are permitted to work the land but do not have the official land-holding rights of an *ejidatario*.

There have been numerous illegal land sales in the *ejido*. In particular, *repobladores* have purchased land. This has created problems in the parcelization process as *repobladores* who have 'informally' purchased *ejido* lands have no formal, legal rights to the land, despite the purchase. The comisariado interviewed at the time, Humberto Domínguez, emphasized that sale of *ejido* land was not legally permitted – stressing that the only legal means of land transfer was through marriage and inheritance. He admitted that '*muchos venderían si pudieran*' ('many would sell if they could'), out of simple economic need. Another Nicolás Bravo farmer, Gilberto agreed that, '*muchos sólo esperan el título para vender y irse. Por eso no quiere dar el gobierno la tierra*' ('many are just waiting for the title to sell and leave. This is why the government doesn't want to give the land'). Not everyone wants to sell, in his own case: '*no pensamos vender, si vendemos – ¿ De qué vamos a trabajar después?*' ('we aren't thinking of selling, if we sell what will we work in afterwards?'). It is relevant to the case study that land speculation around the Programme for the Certification of *Ejido* Land Rights (PROCEDE) privatization process was intense during the early stages of this research. PROCEDE is a government organization charged with the parcelization and titling of communal *ejido* lands, making formerly communal, usufruct lands saleable if communities voted to undergo parcelization. PROCEDE is the result of the repeal of land reform established after the Mexican Revolution of 1910 with the amendment of Article 27 of the 1917 Mexican Constitution in 1992. However, the repeal of land reform and subsequent creation of PROCEDE did not engender a wholesale change in the *ejido* sector, as *ejidos* themselves were ultimately given control over whether lands could be privatized and the rules to which *ejidatarios* must adhere (Perramond 2008).

The issue of migration to the tourist poles (in the mid-1990s and later to the United States) also added a layer of complexity to the land tenure circumstances in Nicolás Bravo. According to a new regulation promulgated by Nicolás Bravo to protect the land rights of *ejidatarios* who work outside of the community, they could leave for 2–3 years and still retain *ejido* rights, providing approval was obtained in advance. However, they would give up lumber extraction income, while they remained outside the community.

The principal cropping system in Nicolás Bravo has long been *milpa*, with around 1,000 hectares planted annually throughout the vast forest areas of the *ejido* at the time of this research. Average yields had declined from around 4 metric tonnes/ha (high by Yucatán Peninsula standards) to 2 tonnes/ha. Local farmers attribute these declines to the land being 'tired' (because of a shortened fallow), poor seed stock, population increases and inadequate transportation thus increasing pressure on land near the village and roads. Despite declining yields, Comisariado Domínguez explained many people make *milpa*, because the little remaining support provided by PROCAMPO (state support programme for subsistence agriculture) is for corn (maize) paying each farmer 500 pesos/ha of *milpa* planted. In addition to *milpa*, the *ejido* had 200–300 hectares of maize *mecanizado* (machine tilled and harvested and using chemical inputs, although rain fed); 525 hectares of irrigated rice cultivated by Belizean Mennonite farmers who rent land; 300 hectares of rain-fed jalapeño chili planted as a cash crop by approximately 150 local farmers and 20 hectares of lemons and oranges planted by three farmers. Also of significance in the *ejido* was honey production, with 90 beekeeping households and 53 *ranchos* (parcels with cattle), with a total of 1,000 head of cattle. With respect to forest extraction, the *ejido* had 20 remaining *chicleros* and an annual timber allotment. Overall, only a small portion of the arable land in the *ejido* was under cultivation.

The farmers of Nicolás Bravo have faced conditions of chronic undercapitalization that mirror the experience of smallholders across Mexico. Although these conditions existed prior to the neoliberal restructuring that commenced in the 1980s – the general consensus is that making a living from the land has become considerably more difficult with the reduction in agrarian subsidies, the elimination/reduction in rural credit (in 1989) and phasing out of CONASUPO federal marketing programmes, among other government divestments associated with neoliberal reforms (Chase 2002; Martin 2005). The lack of capital and infrastructure combined with short-term market gluts have deepened the heavy reliance on outside intermediaries who only pay farmers a pittance for their crops.

This situation in the Yucatán has led to a growing exodus of people from the community in search of work and/or better life opportunities (Carte *et al.* 2010). This is particularly true of the children of *ejidatarios*. During the mid-1990s, approximately 60 *ejidatarios* (or their children), comprising approximately 10 per cent of the *ejido* population, worked in the tourist poles on a temporary basis in construction work or in hotels. Most of these people were young men, many of whom were married. The majority of these temporary migrants continued to make *milpa*, often leaving their children or spouses in charge and/or hiring additional labour as necessary. Most continued to go back and forth between the town and the tourist pole in an attempt to maintain their crop. Ramirez admitted, '*cuando se descuide mucho se afecta*' ('when they become careless, it affects their crop a lot'). Approximately 24 *ejidatarios* had abandoned the *ejido* permanently and lost their land rights. Unless more local sources of employment were generated and farmers were able to work year round, Domínguez said that he expected 'more of the same'.[5] He was correct as the household migration survey in this same community in 2003 revealed that over a quarter of all households sampled had some migration links to the tourist poles, and permanent migration was higher than circular migration (in contrast to earlier), and surprisingly, nearly the same proportion had migrant links to the United States (Table 4.2).

Several study informants felt migration has had a negative impact on local families citing, in particular, numerous cases of men abandoning their partners and children. In some cases, men had established distinct households with other women in the tourist poles, leaving their wives and children to fend for themselves. Thus, there has been an increase in female-headed households in Nicolás Bravo.

Some farmers discovered, however, that they could make a better living in Nicolás Bravo working as full-time growers than by migrating. One farmer, Gilberto, described how he worked in Cancún for 5 years and did not have anything to show for it. He explained, '*me vá mejor aquí … en Cancún gana uno mucho pero gastas todo*' ('I'm doing better here … in Cancún you earn a lot, but you spend everything'). Since having returned to work in Nicolás Bravo as a full-time farmer making *milpa* to feed hogs, keeping bees and planting jalapeño chili peppers, 29-year-old Gilberto has managed to buy a home and establish rights to local land. In contrast, his brother, who recently returned from working in Cancún for 9 years, has nothing to his name. Gilberto concluded, '*No me gusta Cancún … es sólo para los con dinero*! *El que no sabe cree que alli va a tener algo, pero lleva años en Cancún y no tiene nada. La vida es dura en Cancún*! *Aquí se vive mejor*' … ('I don't like Cancún … it's only for those who have money. Those who don't know think that there they will have something but they spend years in Cancún and have nothing. Life is hard in Cancún! Here one lives better' …). Similar comments were reported from San Juan (Momsen 2003).

The following case study of a Cancún *milpero* and his family provides insight into the everyday life of a *milpa* farmer in southern Quintana Roo.

Between two worlds: the story of Don Mario García and his family

Don Mario was selected for this case study not only because of the relationship I had forged with him and several of his adult children whom I interviewed in Cancún but also because he typified the average *milpa* farmer – with subsistence *milpa* cultivation remaining his principal agricultural activity despite continuing attempts to plant cash crops. Don Mario also represents a growing trend within the state towards part-time farming – working in the tourist zone for cash income, while still trying to maintain a viable *milpa* plot. At the outset of the study, he was struggling with the demands of maintaining two homes – one in Cancún and one in Nicolás Bravo, his village. He, personally, expressed a desire to remain in the village and keep his 'hands in the soil', but his wife and older children preferred Cancún.

Don Mario García exists in a fluid world circulating between the two metaphorical *milpas* – the *milpa* of *los antiguos* representing 'Mayaness' and historical continuity and the urban *milpa* of Cancún representing 'progress' and modernity (Re Cruz 1996). As with a transnational migrant, Don Mario has negotiated a complex series of social fields and networks linking him to both Cancún and his village through a continuous circular flow of family members, capital, material goods, information and ideas. The analogy of the transnational migrant applies to the *milpero* immigrant to Cancún, as the social, economic and cultural space between the Cancún tourist space and the Maya rural space is arguably as great as between two nation states (Torres and Momsen 2005a). As with two nations, a complex series of interrelationships between and within rural and urban space, embedded in the broader historical and socio-economic context, induce the flow of migrants between the two realms. Like transnational migrants, rural-to-urban immigrants exist in a hybrid reconstituted social space in which they reconstruct their racial, ethnic and political identities as an adaptation to their fluid, existence and connections linking them to both the urban and rural *milpas*. Maya migrants assume multiple identities as they negotiate their position between and within the two *milpas*. This fluidity not only restructures the life experiences of individuals but also reconfigures family networks and relations – fostering the development of complex extended family networks, which stretch across space, transcending distance and boundaries. This is the context in which Don Mario attempts to negotiate his position between the two *milpas*, struggling to improve his livelihood strategies while distributing risk and keeping his family together.

After farming all his life, 2 years prior to our meeting for the first time in 1996, the 49-year-old Maya farmer, Don Mario García, left his *ejido* (Nicolás Bravo) in southern Quintana Roo. Don Mario had suffered two consecutive years of total crop failure in addition to losing his entire bee population, 60 invaluable hives he had worked years to build up. He attributed the crop losses to a marked change in the climate, combined with a plague of pests, diseases and 'bad luck'. The loss of his entire bee population devastated him. The mysterious 'disease' killed his entire bee stock over night, but oddly did not affect any of his neighbours. Don Mario's children suspected foul play on the part of one of his brothers who was jealous of Don Mario's relative prosperity. This was the same brother who after the tragedy 'helped' Don Mario by procuring his entire bee infrastructure for a pittance, so that he could feed his family. Don Mario and his wife, Doña Elsa, thoroughly depressed by the streak of bad luck the *campo* had dealt them, decided to take their five youngest children and joined their two married daughters living in Cancún (leaving behind one married daughter in Nicolás Bravo). Don Mario and Doña Elsa have a total of eight children ranging in age from 8 to 30. They have managed to make ends meet in Cancún, between Don Mario working on occasional construction jobs and Doña Elsa making and selling *salbutres* and *panuchos*, Yucatecan fast food, in the local markets.

Doña Elsa no longer wears the *huipil* nor do any of her daughters. While in Cancún, the family joined an Evangelical church to which their eldest daughter introduced them. In addition to attending church gatherings several times a week, they regularly attend faith healing sessions and the local campaigns of travelling charismatic evangelists. Although the oldest children, Don Mario and Doña Elsa, speak Maya among themselves, the youngest children, although they understand it, are not fluent in the language.

The family lives in a small shack in a squalid area of the Franja Ejidal. They live without running water, plumbing, paved roads and other basic services. The Franja is a poverty continuum, ranging from older areas near Cancún's downtown, which have become middle class neighbourhoods, to the squalid, squatter camp fringe spreading northward into the adjacent Isla Mujeres *ejido*. At its edge, the Franja consists of nothing more than a grouping of tar paper shacks and crude thin-stick structures topped with patchworks of anything that will keep out the rain. In Nicolás Bravo, their large airy, oval cement floored *palapa* and adjacent cement block building also remain for when they return to the village for visits.

During our first meeting in late November 1996, we met at Don Mario's 29-year-old daughter Raquel's house in the gentrified area of the Franja Ejidal, which was now fully served by all basic infrastructures. Raquel's cement block home, in the middle class section of the Franja, reflected her relative success working as a Hotel Zone domestic worker for various families and her husband Martín's salaried position as a tour bus driver. Although she and her husband had migrated 8 years before and had done well in Cancún, Raquel lamented the decline in her parent's economic circumstances. Raquel recalled better times when her father had a relatively successful farming operation in Nicolás Bravo, where the family never experienced hunger as they often do now. She recalled accompanying her mother daily to the Chetumal market to sell the fruits, vegetables, honey, flowers, chickens and piglets produced by her family.

During the first meeting with Don Mario, it was clear that he was facing a critical decision whether to return to farming or remain permanently in Cancún. He missed farming – his true trade – and life in Cancún had proven difficult with the lack of permanent employment for the kind of low-skilled jobs that he was able to obtain with his limited education. In addition, the living conditions in the shantytown were poor, crowded and plagued by juvenile delinquency, drugs and other social problems. In recent years, the family had lived constantly on the edge of hunger. At the time we met, they only survived with the help of Raquel and her husband, Martín. The most important factor in his decision whether to return to the village, however, was that his *ejido* Nicolás Bravo was initiating the process of land parcelization. Don Mario had been an *ejidatario* since he arrived from Yucatán 26 years ago and he had rights to 100 hectares. The *ejido* assembly had recently voted that only *ejidatarios* who planted and fulfiled the community labour *fagina* requirements would be eligible to receive a parcel and eventual land title. After his extended absence from the village while living in Cancún, Don Mario ran the risk of losing his rights to the land he had farmed for 24 years. Matters were further complicated by the fact that Doña Elsa no longer wanted to return to the village – her 10-year-old son Isidro had received an academic scholarship in Cancún, and she felt that the city was more interesting than Nicolás Bravo. Don Mario only had 1 month to decide whether he should start clearing his land to plant the *milpa*. In January it would be too late. At the time, Don Mario lacked the capital necessary to sow the *milpa*. Ultimately, the family decided that Doña Elsa would remain in Cancún with the children while Don Mario returned to Nicolás Bravo to make *milpa*. Doña Elsa and Don Mario allowed 17-year-old Jesús to drop out of school, as he had long wanted to do, in exchange for his commitment to work with his father and learn how to make *milpa*. Jesús had no previous experience working in the *milpa* as his father had previously managed it on his own with the assistance of paid labour. The eldest children pooled resources to provide Don Mario with transport back to Nicolás Bravo and enough cash to buy seeds and hire a few days of labour during the heaviest forest clearing period.

In early January 1997, Don Mario returned to Nicolás Bravo with Jesús, but they were unable to begin clearing because of a land conflict with two other *ejidatarios*. Normally, family usufruct land rights were recognized in the *ejido* and respected by other *ejidatarios*. The land in question was a piece of *monte alto* (older growth), which had not been farmed since his father had last cultivated it 24 years ago. This entitled Don Mario to usufruct rights to the area. This particular land was valuable not only because of the old forest growth but also because it was flat with deep, fertile, black soils and it was located only 500 metres from the road. With the heightened land speculation in the *ejido*, a direct consequence of the PROCEDE programme that promised future individual titles, there was increased land conflict and pressures. According to Don Mario, *ejidatarios* without a claim were ‘grabbing for land’. There had been so many problems that state agricultural authorities who oversee the parcelization process of PROCEDE became involved and informed the *ejido* that if they could not stop the fighting, there would be no land titles granted to anyone in Nicolás Bravo. Eventually, in mid-January, Don Mario’s land conflict went before the *ejido* Council and they ruled in favour of his rights to the land. The conflict delayed his *tumba* (forest clearing) by 1 month, which meant Don Mario was in a race against time to clear the lots by March, which was necessary to ensure an adequate drying period before the rains in May to permit a '*buena quema*' (a good burn). To make matters worse, Don Mario had only Jesús, who had no prior *milpa* experience, to help him fell the forest and clear the land. He did not have the cash necessary to hire outside help, which had become prohibitively expensive given the competition for labour with the tourist areas.

Don Mario planned to attempt to clear a total of 4 hectares of land – 2 hectares of 24-year-old *monte alto* growth, one of *monte bajo* 10-year-old growth and one of 4-year *huamil*, a low secondary growth with a relatively short fallow. He chose to include the 1 hectare of *huamil*, which was easier to clear but would yield less than older growth, given the time constraints. Don Mario explained that a minimum of 8–9 years of fallow was needed to get a decent yield. He expected to yield up to 1.5 TM/ha in his best older growth areas.[6] He did not plan to sell any of the production but rather acquire 15 hogs that could be fed with production from 2 hectares of *milpa* and sold for cash. Don Mario also planned to plant 1 hectare of *achiote*, a red spice used in Yucatecan cooking, which was highly profitable. He explained that *achiote* was a secure crop because '*no hay mala cosecha y tiene mucho mercado*' ('there is no bad harvest and there is a big market').

Don Mario also considered getting back into honey production and planned to request a soft loan for hives and subsidized equipment from a government sponsored programme, *Alianza para el Campo*. However, time was running short as the honey harvest period was 4 months ranging from January through May. After visiting the *Alianza para el Campo* office in the nearby city of Chetumal in February, Don Mario was disappointed to learn that assistance criteria had become more stringent. To receive the equipment on a credit basis, albeit at a subsidized rate, he had to provide a down payment of 500 pesos – an amount he did not possess. He resigned himself to the fact that he would not be able to enter into bee keeping that year. After learning of this, however, and knowing the potential profitability of bee keeping, his adult children again pooled their resources and sent the necessary funds to Don Mario. Don Mario returned to the office, submitted his paperwork and the deposit. After significant delays, much to his dismay he learned his paperwork had been lost and he had to resubmit. His ability to tend to these bureaucratic exigencies was hindered by the fact that he spent several months in Cancún and was unable to attend to the persistent in-person follow-up necessary to move the paperwork. By the time the bureaucratic processes were completed, it was well past the honey production season.

The remainder of this case study will follow Don Mario and his family through the *milpa* cycle of a Cancún *milpero*. The *milpa* cycle will be used as a temporal framework from which to anchor the discussion, which will weave together both technical crop management information and family history to gain insight into how tourism-driven migration and neoliberal agrarian policy play out at the ground level in the life of one farmer and his family. Using the *milpa* as a temporal frame is appropriate as nearly all the activities for Don Mario and his family during this period were affected by the rhythms of the *milpa* life cycle.

Brechar *and* Tumbar: *the race against time*

Normally, the *milpero* begins to clear the labour-intensive *monte alto* as early as August or September to allow sufficient drying for the burn. Don Mario, who did not begin clearing until February, was extremely behind schedule. He faced a race against time. The delay was related to various factors, including the need to earn cash through occasional labour in Cancún, the desire to be with his family in Cancún, land disputes in his *ejido* and indecision on whether to plant at all. Don Mario had to compensate for the lost time by recruiting family labour and, if at all possible, paid labour. He decided to limit his selection of *monte alto* land, which while more productive would also be more labour intensive. Despite his efforts, however, he would pay a price for his late start.

The first task Don Mario undertook was to create the border to demarcate the planting areas. Land clearing is the most labour-intensive task and the timing is critical. An area that is cleared too soon will experience regrowth, a poor burn and thus subsequently low yields. Areas that are cleared too late will not have sufficient time to dry out for the burn that is normally done in April. Typically, the more labour intensive, older growth is cleared first. The higher plant moisture early in the season makes cutting the larger trees easier. Also, the older growth, having been felled, requires more time to dry out. Don Mario was not able to begin *la tumba* until January, a month later than he had planned, which pushed back his entire cropping schedule. Given this delay, and the lack of family labour (other than inexperienced Jesús), Don Mario acknowledged his plan of 4 hectares was ambitious and perhaps beyond reach.

Don Mario started with the 24-year *monte alto* that had been at the heart of the land conflict, clearing the large trunks first that required more time to dry out. He moved on to the 10-year *monte bajo* plot but then as he ran out of time, he decided to clear the 4-year fallow *huamil* with less material, which would dry more rapidly (but provide less biomass and lower yields). In the end, he ran out of time and was unable to clear the full 4 hectares.

Don Mario and Jesús performed the bulk of the clearing between 11 February 1997 and 10 April 1997. Son-in-law, Martín, also came from Cancún for an 8-day period in January to help them clear land. Don Mario also hired a field labourer, Pedro, for 11 days spread throughout the entire period of clearing. During the last 3 days of the clearing, 15-year-old son, Alberto, who was visiting from Cancún, helped out in the field as well. Both Jesús and Alberto expressed a dislike for the field work and Jesús was already contemplating going back to school. Don Mario managed to clear a total of 80 *mecates* (3.2 hectares) over the 3-month period – almost 1 hectare less than his 4 hectare goal. While Don Mario was still completing the forest clearance, others were burning their fields. His late clearing completion date (10 April) hurt him in two ways. First, it did not permit the ideal 2-month drying out period necessary for a good burn. Without a good burn there are more solid trunks and material remaining and less of the thick ash layer that provides the crop with necessary nutrients. Second, the late clearing completion date forced Don Mario into a game of 'chicken' with Mother Nature. He had to try to hold out as long as possible before burning to permit drying; however, if the rains were to start, he could lose the opportunity to plant at all. If it were to rain prior to the burn, there would need to be a period of at least 15 dry days to permit sufficient drying to initiate a burn. Normally, the hottest month is in April, and this is the safest time for farmers to burn. Upon entering into May, the month in which rains often begin, the farmer is taking a very significant risk. For Don Mario, the stakes were high – having invested his children's resources and his own time and effort in the clearing, he could not afford to fail. Normally, once rains started, it would have been unusual to have a 2-week period without rain – the minimum time necessary for drying prior to burning. Such a result would be disastrous – planting would be impossible and the labour and cash invested a complete loss.

After completion of the clearing, Don Mario and Jesús returned to Cancún during the 1-month drying period. They planned to return to Nicolás Bravo in 30 days to initiate the *quema* (burn) and plant the *milpa.*

La Quema: playing 'chicken' with the **Chaak**[7] *and the 'bad' burn*

As noted earlier, the *milpero* normally burns his *milpa* plot during the hottest days of April when he has the highest probability of a good burn. Don Mario, however, had already decided he must wait until May. Even then, he knew he would have a problem burning the largest felled lumber. As we entered May, Don Mario tried to hold off the burn to the very last minute. He sent word to me that he would conduct the burn sometime between the 5 and 10 May. He was unable to establish an exact date as he required specific conditions on a given day to achieve a successful burn. Don Mario explained that for a good burn, a minimum of 15 days of hot sun without rain were needed, and during the day of the burn, the winds should be steadily blowing in one direction. It is important to have as little green material as possible; everything must be sufficiently dried out with steady winds.

After a visit to his field on 8 May, Don Mario tested the temperature and the wind speed. Finally, after agonizing a couple of hours, he decided not to burn that day as the wind speed was not adequate and the wind direction was inconsistent. Looking up to the sky, he noted the clouds gathering towards the end of the day signalling that the rains were coming. While he acknowledged it was risky to wait, as it could have rained the next day he decided to hold off for stronger, more consistent winds. As long as it did not rain that night, Don Mario felt he could achieve a good burn the next day. All other farmers, excepting a few unfortunates sharing his predicament, had already burned their plots and several had planted. While many viewed the accumulating clouds with hopeful anticipation of rain, for Don Mario, it would have been disastrous.

As Don Mario arrived home from his *milpa* plot late on 8 May (6:00–7:00 pm), the earlier clouds turned black and menacing. The air was cool and damp, signalling that rain was imminent. For one very alarming instant, a light drizzle passed over Nicolás Bravo, but it lasted only a few minutes. The remaining evening and night was one of tension and dread at the prospect of waking up to the sound of rain. The rain held off through the night.

On the morning of 9 May, Don Mario spent the morning hunting for deer in hopes of selling some meat for cash, as Doña Elsa had arrived the previous night from Cancún needing money to buy food and school uniforms for the children. That afternoon, Don Mario decided that it was a sufficiently clear, sunny day with a reasonable wind blowing from the north. The rain scare of the previous evening had convinced him that he could not wait any longer and risk that the rains might begin. At approximately 3:00 pm in the afternoon, he arrived at his *milpa* site and cut two large branches of dyewood, lacerated their ends and used them as torches to ignite the fire. Don Mario explained that to burn he did not require any special equipment or fuel. Everything he needed was right there in his field.

The resulting burn was disappointing, to Don Mario, but not devastating. As a result of weak winds and an inadequate drying period for the most recently cleared portion of the plot, the fire never achieved the desirable intensity. Consequently, a significant amount of unburned material remained in the field as opposed to being integrated into the soil as ash. Don Mario's hopes of planting *achiote* were also dashed, as only corn would grow in an area with such a poor burn. Don Mario considered gathering and attempting to ignite the unburned material, but he decided against this approach as it was labour intensive and the likelihood of sequential days without rain was unlikely. Also, a re-burn would prevent him from planting squash or pumpkins, both of which required a head start before planting the taller corn, which would otherwise shade and inhibit their growth. He decided to plant the entire 3.5 hectares with *chibua* squash before planting the corn, bean and squash combination. While *chibua* can be eaten whole when it is young, it is primarily grown as a cash crop for its seeds that are roasted and commonly sold as a snack food or as the base for the typical Mexican *pipián* sauce. Don Mario was disappointed with his burn, but he had seen worse – it was sufficient to make *milpa*.

***Planting, replanting and weeding: battling the* mapaches**

Don Mario waited 3 days after the burn to allow the soil to cool and then he and Alberto planted the entire 3.5 hectares with the *chibua* squash using self-crafted digging sticks. Two weeks later, after a test planting of corn, *frijol de tallo* and yellow squash mixed (to see if wildlife would consume the seeds) Don Mario, Alberto and Martín (from Cancún) began to plant the remaining crops, including some patches of taro and purple and white sweet potatoes, 12 hours a day over 4 days.

Three days after completing the planting, they noted that animals were digging up the seeds. The primary culprits were the clever *mapaches* (raccoons). In some areas, the animals consumed almost all the seeds planted. Don Mario estimated that they had lost nearly 30–40 per cent of the corn and bean seeds. Nearly all the *chibua* perished because of the combined effects of animals, extreme heat and untimely rain. Despite the menacing clouds during the burn, the rain never arrived as it should have around the middle of May. The increased animal problems were in part a function of the rain, because they could see the holes where the seeds had been planted. With rain, the disturbed terrain would even out. Don Mario would not replant the *chibua* as he had already planted corn that grows faster and would shade the *chibua*. Don Mario was worried that the rains would not come before the entire crop was lost. Although other Maya farmers might conduct a *chachaak*, a Maya ritual or ceremony calling for rains from the rain deity *Chaak*, during this circumstance, Don Mario, who was a converted Evangelical Christian, no longer held to Mayan religious beliefs and so did not subscribe to traditional agricultural rituals. He explained that he does not do the *chachaak* or the *primicia*, '*como no soy Católico*' ('as I am not Catholic').

Eventually it rained, and Don Mario spent 2 days in early June replanting the most damaged areas. The need to replant was a common occurrence given the large number of animals that inhabit the forest, including birds, parrots, spider monkeys, deer, racoons, coatis, badgers and wild boars. Although these animals are pests and a threat to the *milpa*, they also represent food and a potential source of cash. During this period, Don Mario was under pressure to hunt deer and wild boar to generate the cash Doña Elsa needed to take back to Cancún. The three younger children – Magdalena (12 years), Isidro (10 years) and Alejo (8 years) – were staying alone in Cancún. They walked to Raquel's home for meals and she checked in on them after she got home from work.

At the end of May, Don Mario's ancient shotgun yielded one deer and a wild pig that he sold to the local highway truck stop restaurant for cash income. Doña Elsa returned to Cancún with the funds she needed for the uniforms. It is a normal sight to see *milperos* carrying shotguns when returning from their fields. Not only do they use the guns to ward off animals entering the *milpa* but they also hunt for food or to sell meat for cash. Don Mario hunts frequently, and on various field visits, we saw a variety of animals, including monkeys, parrots, *tepesquintles* (a type of large rodent) and wild boars that had been captured.

Don Mario had some small earnings from his wood cooperative during this period. This *ejidatario* group shared the labour and profits from clearing designated areas of *ejido* forest. During his last wood group meeting, they distributed 300 pesos to each member, although the members believed their true share should be 500 pesos each. Don Mario believed that the now departed treasurer stole the funds. Corruption is an ongoing problem with leadership in the *ejidos*. Don Mario recalled how one of the previous *ejido comisariados* stole 100,000 pesos, approximately 10,000 USD at the time, from the *ejido*. He fled, eventually emerging elsewhere as a member of the state legislature. Don Mario claimed the *comisariado* got away with it because the government was involved and he favoured certain *ejidatarios* who served as co-conspirators.

After replanting in June, Don Mario returned to Cancún to be with his family, spending most of the months of July, August and September away from his plot. Normally, during this period, the *milpero* would do one major weeding in July, while frequently visiting or staying at the plot to guard against animals and human pilferage – the latter was particularly a problem during difficult times. Don Mario admitted that even he had resorted to stealing from other fields – but only in times when there was no food for his family. In Don Mario's absence, the weeds proliferated and numerous animals grazed his field, particularly racoons, birds, badgers and wild boars. At one point, the animal problem in Don Mario's field became so pronounced during his absence that his brother, who was maintaining nearby plots, called him in Cancún. Don Mario returned briefly at the end of August to find many bare patches, where animals had dug up seed or early germinating plants had died. His corn had also suffered from an infection of maggots and many ears had developed with only a few kernels of grain inside. The *chibua* cash crop was almost entirely wiped out. During one field visit interview, Don Mario's brother estimated that because of the lack of weeding and animals, Don Mario would lose at least half a tonne/ha of his corn yield. Don Mario returned again from Cancún in October to find that animals had caused further damage. The lack of weeding had exacerbated the problem because animals hid in the weeds. Don Mario and Alberto spent 8 days and 112 person-hours weeding, in hopes that this would reduce the animal problem. They also spent several nights in the *milpa* to ward off intruding animals. Don Mario admitted that his absence had harmed his crop, '*necesito estar allá, si no los animales me acaban la milpa*' ('I need to be there, if not the animals will finish off the *milpa*'). Don Mario's nephew made statements revealing the extended family's disapproval of his absence. He remarked that Don Mario's *milpa* '*está abandonado*' ('it is abandoned'). Don Mario's nephew noted that his uncle had far better land than his family, but his crop would produce less this year. With disdain he suggested that Don Mario preferred hunting to working in his field. He commented that Don Mario's family '*les gusta ser turistas - hay demasiados insectos aqui*' ('they like to be tourists – there are too many insects here'). The nephew hinted at family tensions, noting that Don Mario preferred to live in Nicolás Bravo, but his wife and children wanted to be in Cancún.

Reaping the harvest: the return to the rural milpa

By early October, Don Mario and Jesús, now joined by 15-year-old Alberto, who also dropped out of school, returned to Nicolás Bravo to begin turning down ears of corn. The ears are turned down to prevent rainfall intrusion and protect against birds and insects – while also hastening in-the-field drying. The corn remained in the field, while Don Mario continued a thrice-weekly harvest from October through February. Don Mario estimated the average yield for this crop was approximately 1 metric tonne/ha. He had originally expected to achieve 1.5 metric tonnes/ha, but the late schedule, the poor burn, the seed loss, the lack of weeding and animal damage all contributed to lower yields. Don Mario's brothers, with inferior land resources, had an excellent crop by local standards with an estimated yield of 2 to 3 metric tonnes/ha. Don Mario believed if he had been present on a full-time basis, with his superior land resources, he could have achieved these relatively high yields.

Don Mario was unable to use his corn harvest to raise hogs for cash as he had planned. He lacked the capital to buy the piglets. Additionally, his family having returned from Cancún needed to be fed. He used some of the corn, however, for 14 chickens, most of which his family also consumed. Privately, he lamented his inability, because of the lack of capital and a poor burn, to plant the *achiote* cash crop as he had hoped. Now, the cash would have been very useful indeed.

By December 1997, Doña Elsa, the six youngest children and two older married daughters had all moved back to Nicolás Bravo because of the higher cost of living in Cancún and their unhappiness with the family's separation. Only Raquel and her husband, Martín, remained in Cancún. Both their oldest daughter, Margarita (30 years), with her five children, and Eloisa (22 years), with her two children, moved back to Nicolás Bravo, while their husbands remained in Cancún to work in construction and remit money back to Nicolás Bravo.

Don Mario committed once again to attempt to earn a living as a full-time farmer. His physical presence in the *ejido* also strengthened his claim to land in the parcelization process. Those who were not physically present and who did not participate in the *fagina* (communal work obligation) risked losing their rights to an allotment. Don Mario stood to gain 100 hectares of what is considered relatively good agricultural land. Doña Elsa tried to make the best of being back in the village, which paled in comparison to the bright lights of Cancún. She travelled to Chetumal daily on the public bus to sell 'almost anything' she could in the local market – wild boar, pigs, bananas, tomatoes, flowers and so on. She netted an average of 50 pesos a day (after deducting the 42 pesos transportation costs) buying produce locally and selling it.

The next growing season, Don Mario received his 20 bee boxes on credit from *Alianza para el Campo* and commenced honey production. With respect to *milpa*, he had already cleared 4 hectares and was clearing the last one for a total of 5 hectares. Don Mario had not enrolled in the PROCAMPO programme the prior year, in part, because of his late scheduling and also his disbelief that the programme would actually pay him to make *milpa*. This year he enrolled and received 550 pesos/ha to grow the 5 hectares. He planned to use some of these funds to buy genetically improved seeds which would yield more. Don Mario had also tapped into government programmes that he had never before explored because of his lack of trust in government programmes and his dislike and fear of credit arrangements.

Despite their decision to return to the village, Don Mario and his children continued to flow back and forth within the social field linking Cancún and Nicolás Bravo over the years. To this day, they continue to maintain strong economic, social, cultural and familial linkages to both the urban *milpa* of Cancún and the *milpa* of *los ancestros* in Nicolás Bravo, negotiating their position between the two *milpas* in attempts to maximize their livelihood strategies and spread risk across space. Should one *milpa* not yield much, the other remains as a safety net. The social field has extended to a third distant *milpa* 'al norte' when in 2005, Mateo, Don Mario's grandson migrated to North Carolina and later to Arizona working as a dishwasher and remitting funds back to his parents and younger siblings. His father had been saving to pay the coyote costs to cross the border to join Mateo in the United States.

Throughout this case study, Don Mario's *milpa* practises have been indirectly contrasted to those of a 'normal' or 'typical' *milpa*, but the truth is that there is no one reality and no single true *milpa*. The wider socio-economic context in which *milpa* production is embedded and the disparate life circumstances of the modern *milpero* are such that a diversity of *milpa* systems and family survival strategies have evolved in Quintana Roo. The experience of Don Mario, a Cancún *milpero*, and his family, however, is one increasingly shared by farmers across the Yucatán peninsula as they traverse the two '*milpas*'.

Conclusions

Neoliberal reforms have clearly left their mark on the Quintana Roo countryside. The combination of the adoption of PTD as essentially the sole vehicle for economic development in the state, along with the implementation of neoliberal agrarian reforms contributed to the further decapitalization of the countryside and created conditions ripe for mass rural-to-urban migration. During the early stages of the resort's development, Cancún attracted rural workers as an inexpensive way to build the resort and serve the continuing needs of tourists and urban residents. Migration patterns are not uniform across Quintana Roo – with some communities such as Morocoy exhibiting high levels of emigration, while other communities such as Tixcacal de la Guardia (a Maya ceremonial centre) experiencing virtually no migration. The tapestry of migration is not only diverse but also constantly evolving – with circular migration shifting to permanent settlement in many communities and the rapid emergence of migration to the United States over the past decade. Given the heavy construction and thus jobs of tourism resorts along the Riviera Maya during this same period, it would seem unusual that so many decided to migrate '*más al norte*' given the high costs and risks associated with international migration and the general absence within the area of migratory experience, networks and ties to the United States. However, reports of mass migration of impoverished indigenous people from Chiapas, and the existence of exploitative worker camps in Cancún, suggest that the insatiable demand for ever cheaper labour inherent in neoliberal development continues to draw poorer farmers (many victims of the North American Free Trade Agreement and other neoliberal agrarian policies) from farther a field (Zunino 2008). Quintana Roo rural residents explained that wages in the resorts are low and competition for jobs is great, despite the continued coastal development, thus making migration to the United States necessary. It is also possible that through work in Cancún, workers do gain knowledge and skills such as English, which facilitate the further step of migration to the United States.

Despite these new migratory trends, in several communities, farmers, such as Don Mario, choose circular migration over permanent migration for a variety of reasons. In Don Mario's case, he needed to maintain some presence in the *ejido* to retain his land rights. Also, with his limited education, he is relegated to lower paying seasonal work involving heavy physical labour in the tourist areas. Indeed, in the tourist resorts, the better permanent jobs are typically occupied by the more educated and more 'sophisticated' people with better Spanish language skills – most of whom come from other states and particularly cities such as Mexico City and Monterrey (and are notably not indigenous people). *Milpa*, while not profitable, did provide a safety net for Don Mario to ensure that he could feed his family. Similar to what Re Cruz (1996) found with the Cancún Maya from Chan Kom, Don Mario also expressed a desire to maintain a connection to his land through making *milpa*, and to his Maya identity, because 'making *milpa* is what *Mayeros* do'. Thus, farmers such as Don Mario live circulating between the two *milpas* for sustenance – Cancún for wage labour and the campo for corn.

Tourism-driven migration has also had important impacts on agricultural production in the Yucatán. Although some farmers have completely abandoned agriculture, many continue to farm as absentee or part-time growers. Typically, these partial-absentee farmers rely more on paid labour to do the heavy work. However, as a result of the competition for labour between tourism and agriculture – agricultural wages have become prohibitively high for farmers. As such, the considerable wage gap between workers in the tourist resorts and agricultural workers makes it difficult for farmers to recruit labour (and the availability of family labour is decreasing with emigration).

Don Mario's case study illustrates the disruptive nature of circular migration to the 'other' *milpa*, Cancún, not only because of the lack of labour and supervision and the 'short cuts' but also because of the critical issues of timing. Although milpa farming is commonly portrayed as simple, indeed, as any other farming system, it has its temporal complexity, which is embedded in locally specific and often rapidly changing environmental conditions. The delays he experienced in clearing and burning his field resulted in lower yields. Additionally, farmers must adapt, be flexible and make quick decisions, all of which are difficult if he is physically removed from the field and cannot observe the crop directly.

Tourism-driven migration, although similar to other forms of labour-driven movement, does possess peculiarities that merit study as a unique form of migration. The seasonal nature of tourism translates into fewer employment opportunities on a year round basis. In addition, the seasonal nature of agriculture, where periods of peak labour at times coincide with the tourist high season, bring the two sectors into direct competition. Tourism, as a service-based industry, draws workers into the tourist space where impressionable rural youth are exposed to displays of wealth, conspicuous consumption, heavy advertising, the leisure culture and foreign values and belief system among other things. Although this occurs in many rural-to-urban migration contexts, in the case of mass tourism (particularly in Cancún, known sarcastically among locals as 'Gringolandia'), the materialism, hedonism, pleasure-seeking and consumption are often exaggerated and excessive in the artificial tourist space unlike in any other context. Also, the tourism industry is extremely sensitive to the whims and vagaries of international tourist preferences – thus leaving immigrant workers even more vulnerable to global shifts in tourism. This is further exacerbated by the fact that many regions that rely heavily on tourism, such as Cancún, lack diversification in their economies – again leaving immigrants dependent upon a monoculture economy.

The wider socio-economic context in which *milpa* production is embedded and the disparate life circumstances of the modern *milpero* are such that a diversity of *milpa* systems and family survival strategies have evolved in Quintana Roo. The experience of Don Mario, a Cancún *milpero*, and his family, who live and work between the two *milpas*, is one increasingly shared by farmers across the Yucatán peninsula.

Acknowledgements

This research was made possible through the generous support of an NSF Early CAREER Award (#0547725); The University of Texas at Austin (UT) Harrington Faculty Fellowship; UT College of Liberal Arts Faculty Fellowship and East Carolina University (ECU) College of Arts and Sciences Research/Creative Activity Grant and College Research Award. I am extremely grateful to Paul Skillicorn who accompanied me through much of this field work over the years, serving as a colleague, friend, sounding board, driver, photographer, videographer and babysitter. I thank him for reading an earlier draft of this chapter and for his always insightful feedback. I appreciate the diligent assistance of my research assistant Lindsey Carte, who made considerable improvements to earlier drafts of this chapter. My deep thanks to Don Mario and his family who shared with me their knowledge, experience, fears, hopes and dreams.

Notes

1 The Franja Ejidal is an informal settlement on the periphery of urban Cancún that houses thousands of people, including recent rural immigrants who have arrived to find work in the tourism industry.
2 I use 'to make *milpa*' rather than 'to plant *milpa*' to more closely reflect the expression in Spanish, which is *hacer milpa. Milpa* is not only a farming system but also an important social construct among the Maya. This is in accord with previous ethnographers' use of the term. See, for example, Redfield *et al.* (1934).
3 *Don* is a title of respect, similar to 'Mr' used for men in Latin America. The female equivalent is *Doña*.
4 A *chiclero* is an agricultural labourer who works with *chicle*.
5 During the period of this case study, in 1996–1998, migration to the United States was virtually non-existent in Nicolás Bravo. By 2003, when I conducted a brief household survey of migration which confirmed that 25 per cent of households had at least one member in the United States – and 27 per cent had a member in Cancún or another tourist pole (see Table 4.2). Clearly, during this period, migration also has transitioned from a temporary circular flow to one of permanency with family members remaining in the poles. The shift in focus from internal to international migration is likely related to several factors, including the influx of low-wage immigrant workers from Chiapas and other parts of the peninsula to the tourist poles and possibly a step migration effect related to new skills, experience, consumption patterns and so on acquired in Cancún which facilitate emigration '*mas al norte*'.
6 By US standards, where the average yield during this same period was 8 TM/ha, this yield of 1.5 TM/ha is very low. However, relative to other *milpa* yields in the Yucatán that we recorded, some as low as 100 kg/ha, this was considered a good yield.
7 *Chaak* is the Mayan rain deity.

References

Arnaiz, S.M. and Dachary, A. (1992) '*Cancún: El enclave turístico y sus costos. Subtema: Los Impactos Sociales y Económicos del Turismo*' [Cancún: The tourist enclave and costs. Sub-topic: Social and economic impacts of tourism], Caribbean Studies Association XVII Annual Conference, Grenada, West Indies.

Bever, S.W. (2002) 'Migration and the transformation of gender roles and hierarchies in Yucatan', *Urban Anthropology & Studies of Cultural Systems & World Economic Development*, 31(2): 199–230.

Castellanos, M.B. (2003) *Gustos and gender: Yucatec Maya migration to the Mexican Riviera*, PhD Dissertation. University of Michigan.

Castellanos, M.B. (2007) 'Adolescent migration to Cancún – Reconfiguring Maya households and gender relations in Mexico's Yucatan Peninsula', *Frontiers: a Journal of Women Studies*, 28(3): 1–27.

Carte, L., McWatters, M., Daley, E., and Torres, R.M. (2010) 'Experiencing Agricultural Failure: Internal Migration, Tourism and Local Perceptions of Regional Change in the Yucatan', *Geoforum*, 41(5): 700–710.

Chase, J. (2002) 'Introduction: The spaces of neoliberalism in Latin America', In: J. Chase (ed.) *The spaces of neoliberalism: Land, place and family in Latin America*, Bloomfield, CT: Kumarian Press: 1–21.

Dufresne, L. (1995) *'Migration of Maya peasants to Cancún: Peasantry destructuration and partial urban integration', Actas latinoamericanas de Varsovia: Migraciones laborales en America Latina y su impacto en los lugares de origen y destino*, Universidad de Varsovia, Warszawa, 17: 139–75.

Fischer, P.L. (2007) 'Yucatán as an emerging migrant-sending region', In W. Cornelius, D. Fitzgerald, and P.L. Fischer (eds) *Mayan journeys: The new migration from Yucatán to the United States*, Center for Comparative Immigration Studies, San Diego, CA: UCSD: 1–28.

Fraga, J.E. (1992) 'La migración hacia la franja turistica del Caribe mexicano: Sobrevivencia o superación' [Migration to the Mexican Caribbean tourist zone: Survival and opportunity], *Revistas de la Universidad Autónoma de Yucatán*, 183: 55–9.

Instituto Nacional de Estadistica y Geografía (INEGI) (2005) 'Prinicpales resultados por localidad 2005, Quintana Roo', Conteo de población en la localidades 2000-2005. Online. Available at: www.inegi.org.mx/est/contenidos/espanol/sistemas/conteo2005/localidad/iter/default.asp?s=est&;c=10395 (accessed 1 June 2010).

Kintz, E. (1998) 'The Yucatec Maya frontier and Maya women: Tenacity of tradition and tragedy of transformation', *Sex Roles*, 39(7/8): 589–601.

Martin, P. (2005) 'Comparative topographies of neoliberalism in Mexico', *Environment and Planning A*, 37(2): 203–20.

Momsen, J.H. (2003) 'Participatory development and indigenous communities in the Mexican Caribbean', In J. Pugh and R. Potter (eds) *Participatory and communicative planning in the Caribbean*, Aldershot, UK: Ashgate: 155–72.

Perramond, E. (2008) 'The rise, fall, and reconfiguration of the Mexican ejido', *Geographical Review*, 98(3): 356–71.

Pi-Sunyer, O. and Thomas, B.R. (1997) 'Tourism, environmentalism, and cultural survival in Quintana Roo', In B.R. Johnson (ed.) *Life and deathmatters: Human rights and the environment at the end of the millennium*, Walnut Creek, CA: AltaMira Press: 187–212.

Radel, C. and Schmook, B. (2008) 'Male transnational migration and its linkages to land-use change in a southern Campeche Ejido', *Journal of Latin American Geography*, 7(2): 59–84.

Re Cruz, A. (1996) *The Two milpas of Chan Kom: Scenarios of a Maya village life*, Albany, NY: SUNY Press.

Re Cruz, A. (1998) 'Maya Women, Gender Dynamics, and Modes of Production', *Sex Roles, A Journal of Research*, 39(7/8): 573–87.

Re Cruz, A. (2003) 'Milpa as an ideological weapon: Tourism and Maya migration to Cancún', *Ethnohistory*, 50(3): 489–502.

Redfield, R. and Alfonso, V.R. (1934) *Chan Kom: A Maya Village*. Washington, DC: Carnegie Institution of Washington, Publication No. 448.

Rodríguez, A., Wittlinger, J. and Manzanero Rodríguez, L. (2007) 'The Interface between internal and international migration', In W. Cornelius, D. Fitzgerald, and P.L. Fischer (eds) *Mayan journeys: The new migration from Yucatán to the United States*, Center for Comparative Immigration Studies, San Diego, CA: UCSD: 73–90.

Schmook, B. and Radel, C. (2008) 'International labor migration from a tropical development frontier: Globalizing households and an incipient forest transition', *Human Ecology*, 36(6): 891–908.

Tolen, R. (1998) 'Reinventing ethnicity in Yucatán', *Sex Roles*, 39(7/8): 643–50.

Torres, R.M. (2003) 'Linkages between tourism and agriculture in Mexico', *Annals of Tourism Research*, 30(3): 546–66.

Torres, R.M. and Momsen, J. (2005a) 'Gringolandia: The construction of a new tourist space in Mexico', *Annals of the Association of American Geographers*, 95(2): 314–35.

Torres, R. and Momsen, J. (2005b) 'Planned tourism development in Quintana Roo, Mexico: Engine for regional development or prescription for inequitable growth?', *Current Issues in Tourism*, 8(4): 259–85.

Zunino, M. (2008) 'El Caribe mexicano: Paraíso para algunos, infierno para muchos', Centro de Investigaciones Económicas y Políticas de Acción Comunitaria, Boletín 'Chiapas al Día' No. 567. Online. Available at: www.ciepac.org/boletines/chiapasal-dia.php (accessed 11 May 2009).

5 Female empowerment through agritourism in rural Japan

Atsuko Hashimoto and David Telfer

Introduction

Tourism is often adopted in rural environments to generate additional sources of income by attracting tourists to a rural economy. In some cases, it is an official strategy adopted and promoted by governments to renew declining economies in rural areas. This chapter will examine how the introduction of tourism initiatives in rural Oita, Japan, has provided new opportunities for females' participation in the local economy. In Oita, females have taken a leading role in developing small-scale farmers' markets that are attracting tourists and establishing farm Bed and Breakfast (B&B) inns. Although these initiatives have provided new sources of income, they have also come with additional challenges. Females are expected to continue to practice traditional roles on the farm, thereby increasing their overall work load. This chapter examines to what extent agritourism has empowered females living and working in a traditional society with well-established gender roles passed down from generation to generation. The chapter will begin by highlighting the literature on tourism, agriculture and empowerment and follow with a closer look at the site and the methods used in this study. The findings are then presented in terms of the main themes that have emerged from the research.

Tourism, agriculture and empowerment

Agriculture provides not only the tourism industry resources for food consumption but also the background for attractions in rural environments (Telfer and Wall 1996). Rural areas are often seen as places where busy city dwellers can escape to open spaces with an imagined slower paced, community-based country life. In areas where agriculture has gone into decline, tourism is being utilized as a regional development strategy, with community-based agritourism furthermore helping to promote community development (Telfer 2000; Telfer and Sharpley 2008). Several studies have examined the possibility of tourism increasing the demand for local food products, thereby increasingly the economic multiplier effect (Telfer and Wall 1996; Torres 2003). One of the challenges in sourcing local food is to institutionalize commitments that go beyond the interests and involvement of specific individuals (Telfer and Wall 1996). Torres and Momsen (2004) explored the challenges of linking tourism and agriculture to achieve pro-poor tourism objectives and argue that only through an integrated approach considering all areas of production, producer organization, post-harvest handling infrastructure and marketing, all linked through strong strategic alliances, is it possible to create sustainable linkages between the two sectors.

An emerging area of study and the focus of this chapter is the importance of female empowerment, as females go through the transition from agriculture to tourism-related employment. Female employment in tourism can influence gender roles (Kinnaird and Hall 1994; Momsen 2004). In very traditional societies, the shift to more females engaged in tourism-related work has the possibility of empowering females with new roles outside their traditional positions in society, but some communities are more open to such changes than others. In a study on Belizean women and tourism, Gentry (2007) found that tourism employment is exploiting traditional beliefs, particularly in reference to sex-based segregation, the double work day, housewifization of labour, low wages and low education levels. However, with respect to household economic decision making, business ownership and levels of autonomy and social interaction, tourism employment was viewed to be challenging the norm.

Costa (2005) notes that in rural Greece, females face an increasingly complex situation as their economic roles expand through tourism, while they must simultaneously respond to the expectations of society, propriety and gendered subordination. The females are still expected to fulfil their roles in both production and consumption. One area of transition towards tourism-related employment is the establishment of guest houses on farms. Momsen (2004) suggests that foreign visitors to guest houses and farms often have most contact with female hosts, who provide meals and local guidance, and that this provides welcome social interaction for females in isolated areas. This chapter looks at the changing roles of females on the farm in a developing tourism industry and brings together issues of female empowerment through social interaction, networking and increased economic power in the tourism context.

Japan is one of the most westernized and economically advanced nations in Asia; however, females' station in society has not changed as much as in many Euro-American nations, especially in peripheral areas. In 1999, two important laws were passed in Japan: the 'Basic Law for the Society of Male–Female Joint Participation', or Gender Equity Law, and the 'Basic Law for Food, Agriculture and Farming Communities'. These laws led to a set of 'Recommendations for promoting gender equity in the farming, forestry and fishing communities' by the Ministry of Agriculture, Forestry and Fisheries in Japan (MAFF 2007). The laws were meant to create a new dynamic of responsibility sharing and labour in rural life, encouraging females to be more actively involved in entrepreneurship. There are, however, criticisms that the gender equity law only 'recognizes' the often unappreciated duties and responsibilities that females have long been bearing and that by bringing these tasks into the spotlight, social expectations for females to be more responsible may increase (Miyagi 2001). In addition,

Table 5.1 International comparison of people older than 60 who are currently working an income-generating job, 1995

Age group	*Percentage of age group in work force*		
	Japan	*United States*	*Germany*
60–64	63.9	44.3	17.1
65–69	44.8	29.4	7.5
70–74	38.3	18.7	1.1
75–79	20.6	7.7	0.9
80+	13.7	2.2	1.3

Source: Takahashi (2002).

the population in the peripheral rural areas is ageing. Among economically developed nations, the percentages of people in the age bracket of 70–74 are Japan 38 per cent, the United States 18.7 per cent and Germany 1.1 per cent (Takahashi 2002, see Table 5.1). The MAFF 2007 White Paper revealed that over 50 per cent of the full-time farmers in Japan are more than 65 years old (Excite News 2008). Other statistics illustrate that between the ages of 50 and 74, full-time female farmers outnumber male farmers (Honkawa 2008). In the research area which will be further described in the next section, male life expectancy is 77 and female life expectancy is 84 years; however, males who reached the age of 80 are expected to live another 7 years and females at 80 years of age are expected to live another 10 years (Oita Prefecture 2009). Various levels of governing bodies are aware of the importance of not only educating females about 'empowerment' but also educating male members of the family to understand the significance of female empowerment (e.g. Kumamoto Prefecture 2003).

Background of the site

The focus of this study is the Kunisaki Peninsula, located in the Prefecture of Oita on the island of Kyushu (Figure 5.1). This is a rural agricultural area of Japan with many historical attractions. There are several key issues for the region which have an impact on this research, including an ageing population, depopulation and smaller communities being grouped together for administrative purposes. For example, Matama town, the focus of the research, became part of Bungotakada city in 2005. With both an ageing population and people leaving the region, the farming sector faces some serious challenges. Many of the people who were part of this study were in their 70s and still working full time as farmers. One of the other main issues in this research is the need to understand the traditional attitudes held for centuries with respect to gender roles in society. Males have traditionally been the head of the household, with legal documents in the name of the husband. Although work on the farm is typically shared equally between husband and wife, domestic responsibilities fall mostly on the females. The communities in the mountainous peninsula also have a long history of

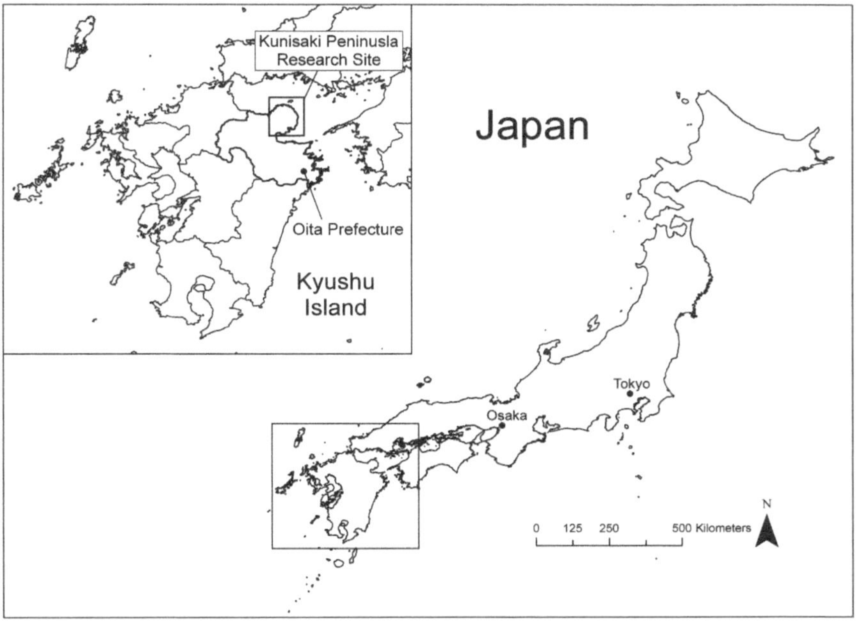

Figure 5.1 Map of Japan showing Kyushu Island and the Kunisaki Peninsula (map by Ingrid Haekel, 2010).

geographical isolation due to the lack of road networks and strong in-group kinship, which together have caused the communities to look inward rather than outward. It is important to keep in mind that this is very different from more urban centres, where attitudes and roles are changing at a much greater speed.

The area under investigation has adopted some of the strategies associated with 'Green Tourism' with some support from the local government. In contrast to the ecological definitions of Green Tourism in the West, Green Tourism in Japan, as defined by MAFF, refers to vacation activities in farming, fishing and/or forestry villages. Examples of Green Tourism activities promoted in the study area include a landscape museum, farm stay B&Bs, and farmers' markets. All these types of activities are investigated in this study.

Research methods

This chapter is based on ethnographic research conducted by the authors, who have travelled to the region multiple times (2003–2004, 2005, 2007 and 2008) in an attempt to gain the trust of the local community. Even though the first author is Japanese, she is still considered an outsider, as she is not from the region under investigation. Both authors participated in daily activities related to agriculture, including planting and harvesting and delivering vegetables to the local farmers' markets and supermarkets. The authors also participated in community events such as rice planting and *soba* (buckwheat) noodle making and attended organic farming seminars. Beyond participating in everyday life related to farming, the authors used a number of research methods to investigate female empowerment, including site visits, key informant interviews and focus groups. Focus groups were conducted with female farmers and also with females making processed food products using locally harvested food to be sold at local markets. By returning to the research sites frequently, the community members grew more accustomed to the presence of the researchers, which allowed the authors to rely more on conversational interview techniques.

Findings

The findings below are arranged around the following main themes that emerged out of the research: (1) new sources of income – new sense of independence, (2) conflicts of traditional gender roles, (3) challenges in meeting the needs of tourists, (4) conflicts in time management and (5) socializing.

New sources of income – new sense of independence

The new tourism projects, such as running a B&B or supplying food to the local farmers' markets, have brought new sources of income to the female farmers. Both the B&Bs and farmers' markets are visited mainly by domestic tourists. In a typical rural household, the family uses only one bank account under the husband's name and the husband has signing authority. However, female farmers participating in Green Tourism projects open their own bank accounts in which to deposit their earnings. For these new accounts, the females have sole signing authority and they are the only ones who can withdraw money. The Japan Agriculture Association's *Female Empowerment Section* considers a female with her own bank account as the initial step towards female empowerment and financial independence. The association stresses that having a personal bank account encourages female farmers to have:

1 an understanding of balance sheets (income and expenses),
2 a higher work morale and
3 self-recognition as entrepreneurs.

Goods sold by females include surplus produce from their fields (e.g. seasonal vegetables, fruit, flowers, herbs and seedlings), fresh seafood (e.g. seasonal fish, seashells and sea weed), homemade food products and crafts, often made during the off season. Additionally, females can earn an hourly salary by working at the farmers' market, shops, or at one of the other agricultural attractions in the region. One group of females was also preparing traditional Japanese lunch boxes (*bento*) using local ingredients to be sold at an area heritage tourism district. Although the income is not substantial, it is enough to cover part of the household expenses and even for humble indulgences that females never had a chance to partake in before. What is more, having a personal bank account provides females with a sense of financial security and independence that they do not otherwise have with a single household bank account in their husbands' name.

Conflicts of traditional gender roles in rural life

While investigating female empowerment through agritourism in rural Oita, one of the main issues uncovered is that many traditional values are still deeply held with respect to gender roles. One of the projects developed under the Green Tourism umbrella is the establishment of local farmers' markets. The farmers sell their best-quality produce through the Japan Agriculture distribution system. The local farmers' markets specialize in selling second-class quality products that would not traditionally be accepted for retail through the Japan Agriculture distribution system. Many of the markets are stopping points for tourists. In Matama, funding for the market buildings came from the local government. Although on paper the official head of the market is a male, he was never actually seen working in the market and it was the females who ran the farmers' market on a day-to-day basis. The females in charge of daily market operation were the people communicating with local government officials regarding start-up funds. Dealing with females seemed to be a new situation for the government office. Interestingly enough, though, when asked for further contacts related to Green Tourism initiatives, the local government often provided female contacts, as females are primarily associated with the projects. A related conflicting issue is the cultural attitude towards risk taking. There is a saying in Japan that 'the nail that sticks out gets pounded back in'. This implies that there is risk in taking chances and that the collective best interest must be maintained. Although it is often female farmers who are embarking on Green Tourism initiatives, they do not want to be perceived as great risk takers and therefore await government action for any major decisions.

A focus group of female farmers was set up to meet at the local community hall to investigate their role in and perceptions of Green Tourism. Just before the authors were to leave for the meeting, they received a phone call from one of the female farmers informing them that two males would be present. As mentioned above, these males on paper were the head of the farmers' market project; however, their actual participation in the project was very limited. As we were going to ask questions about the project, they felt it necessary to be present. With the males being in attendance, we were instructed not to ask certain questions that might cause the males to lose face or be embarrassed. The meeting went ahead as scheduled; nonetheless, after the focus group, multiple individual meetings had to be scheduled with the females to ask the questions we were told to avoid. Another example of persisting traditional gender roles came up while conducting a second focus group with females who were making *kakimochi* (deep-fried thin rice cake) to be sold at the farmers' market. One of the females had to leave to go home and make her husband's lunch. She saw it as her duty to make her husband's lunch and thought he would not be able to make it for himself.

Similarly, a B&B operator admitted that while running the B&B, she was still expected to perform her normal domestic chores and participate in farm work. A second B&B operator indicated that she was required to take a number of courses (e.g. health and safety, and hygiene and kitchen standards) from the local government in addition to managing bookings, preparation, cooking and cleaning for the guests. Although her husband is supportive and willing to deal with the staying guests, he did not want to help out on the operational side of B&B business or with domestic chores.

According to an interview with the Japan Agriculture Association's Women's Empowerment Section, farmers older than 50 years are accustomed to the traditional male–female division of labour. Information from the interview revealed that the husband is the sole decision maker on the farm, yet the females' contribution to farming life accounts for above 60 per cent of farm work, not to mention that there is an expectation for females to be nearly totally responsible for domestic affairs. Although males' contribution to the production process is typically greater than that of females, processing of the harvested produce into sellable 'products' along with packaging and delivery is mainly done by females.

The following quotations illustrate a regular day for a female in her early 70s, who is the head of the market on a day-to-day basis and is usually present when the market is open:

> I usually get up about 3 am, get myself dressed, and then cook a breakfast for my husband. He gets out of bed about 4 am and has his breakfast. We leave home just before 5 am to the field. During the peak season, we should be in the field by 4 am. I work at the field until 7:30 am, then leave for the Farmers' Market. The Farmers' Market opens at 8:30 am to the public. I will be working there until 2:30 pm. I will then go straight from the Farmers' Market to the field and help my husband. We work until dusk, then go home. While my husband is taking a bath, I cook supper. We have a quick supper and continue our work in the shed until about 9 pm. Then we retire for the day.
> If it is a hot summer day, we start working before 4 am. I leave for the Farmers' Market at 7 am. My husband will go home about 9 am when it gets too hot to work in the field, does chores he can do at home, and then takes a nap after lunch. We go back to the field as soon as the heat becomes bearable in the afternoon and work until it gets dark (about 8:00 pm in summer). We work in the shed after supper whenever and as long as is necessary.

The same female also told us about a time when her husband was hospitalized for hernia surgery in the summer of 2008:

> My husband was hospitalized for his hernia – he had had this problem for a long time, but finally it became so bad that he needed surgery. While he was in the hospital, the pre-operation period, the surgery and recovery period

> lasted for about two weeks. I had to cover my husband's chores and it was the busiest season for our Gladioli shipping. I had to call my son back from the city, ask my friend to tend the Farmers' Market, and ask my sister to come from a nearby community and help in the evenings. My colleagues who are also in flower farming helped me with various chores. My neighbors and friends also brought meals to us so that we could concentrate on the various jobs. I started working about 4 am and worked until midnight or even after midnight. Some days I had just a couple of hours of sleep. Cleaning the Gladioli was not really my chore so it took me much longer than my husband. You see, flower selection was my chore. Often, I would forget to eat, but my colleagues and friends supported me and reminded me to watch my own health.

To the question of 'what if she fell ill', she laughed and responded that she would still do the same, because her husband does not know how to cook or how to tend the farmers' market.

As can be seen from these two interviews, the female not only works at the market but is also required to continue working on the farm, complete domestic chores and cook for her husband. Although the farmers' market has given her a new sense of independence, additional income and a new forum for socializing, she is also working extremely hard in the process, crossing between both tourism and agricultural sectors. The female farmers repeatedly mentioned that without the family's understanding and consent, especially of their husband, they cannot participate in Green Tourism projects. Husbands show understanding so long as their wives' participation in Green Tourism projects does not affect their domestic responsibilities and fieldwork.

Conflicts in time management

A repeatedly cited challenge to participation in Green Tourism projects is the coincidental timing of the peak farming season and holiday seasons. Although different varieties of produce have respective 'peak' seasons, the farmers' busiest season coincides with school and national holidays. Peak holidays include school spring holiday in March, Golden Week holiday from the end of April to the first week of May, school summer holiday in July and August and school winter break from the end of December to the first week of January. When the field work becomes intense, the B&B operators and female farmers do not have any more time to spare for visitors in the area, as they are not full-time B&B operators or market employees. From the interviews and observations, farmers expressed the following concerns:

1. Visitors from urban areas know nothing about the farming calendar (almanac) and they just visit the area whenever they can. During the peak seasons, farmers do not want strangers near their fields for fear the visitors might mess up the field.

2 The visitors expect their hosts to entertain them after supper. However, the farmers need to go to bed early so that they can start the day early.
3 Some visitors do not see the difference between a hotel in the city and a farm stay (B&B) and want to stay up late, get up late, have a late breakfast and some of them want to stay in their room all day.
4 There is a conflict between having short stay (overnight) visitors and having longer stay visitors. Most of the visitors do not stay for more than 2 nights/3 days.

In addition to the major conflicts between the tourist season and farming peak seasons, female farmers also have to balance the demands from family members as illustrated earlier.

Challenges in meeting the needs of tourists

An interview with a new B&B operator illustrated some of the challenges she faces. New to the industry, she was not used to coming up with new menu items for the guests who wanted to stay longer than 2 days. The recent movement of 'eat local' has been favourably received by farm tourists; however, the tourists from urban areas have very little idea about the limitation in varieties of produce that this movement implies, especially under the 'One Village, One Product' farming schemes. Villages in the area are specialized in designated specific types of products ('One Product') for each season; in turn, the local people have to live with the same produce throughout the season. Those farmers and B&B operators who have little time to develop their culinary skills will have a problem with limited types of available produce to create distinctive menu items for each meal.

Females also faced some challenges in marketing. Leaders of Female's Empowerment in the Japan Agriculture Association and representatives from the Agriculture Promotion Section of the municipal office both agree that the lack of funding and human resources for marketing is a major concern for Green Tourism. Currently, Green Tourism projects are marketed through municipal websites and prefectural Green Tourism websites. However, the farmers involved in Green Tourism are mostly elderly and have little knowledge or interest in website creation or learning marketing techniques. Therefore, website updating is totally dependent on assistance from the municipal or provincial office. Another challenge in marketing is that after the initial 'set-up' period, technical and financial support from the local government is often terminated. The farmers, mainly females, are then left to their discretion to decide how to promote and market their products and activities to the potential tourists.

Green Tourism in Kunisaki Peninsula has been promoted to schools in the nearby cities, and B&B establishments have received school trips (from grades 7 to 12). Although most of the B&B operators welcome school trips that usually fill the establishments (usually about six people per night), the schools are free to change their destinations and the business is not guaranteed. The operators interviewed explicitly express their enjoyment at interacting with the young visitors and giving the students opportunities to learn about rural life, and they enjoy even more the return of these students with their families. Yet entertaining repeat visitors is another challenge for the Green Tourism participants as there are limited tourist attractions (be it natural or heritage) to draw these visitors back again in the future.

Socializing and networking among females

One of the main benefits for females of involvement in Green Tourism projects is time for socializing with other females. At the farmers' markets, females are able to meet up briefly while they drop off their goods before returning to the fields to work. One of the focus groups was held with female farmers as they worked preparing local food products. They were able to chat while they worked and this gave them a new sense of community and belonging with the other females. Again, with the ageing population, most of the females in this group were in their 70s.

Although participating in Green Tourism provides opportunities for socializing, enforcing a support system and strengthening existing friendships, the activities do not go beyond the personal or immediate group level to include groups of females in similar projects from neighbouring villages. As illustrated in the husband's hospitalization case, a female's support system in time of need expanded beyond her immediate family members to networks established through the Green Tourism projects. However, when two managing females of two separate farmers' markets were asked about collaboration or networking with other farmers' markets, both of them answered that there was no communication or collaboration with other markets and that they saw no need for such networking in the future. These two markets were in neighbouring communities separated only by hills but with tunnels connecting them yet the different groups of females did not collaborate. Female empowerment should also encourage the creation of new business-related and personal networks beyond existing relationships. In the case of communities in the Kunisaki Peninsula, however, the historical regional segregation and in-group kinships prevent female farmers from expanding their networks.

Conclusion

Although new developments in rural tourism are presenting new opportunities for female farmers, these must be kept in perspective with the other demands in the lives of female farmers. New sources of income have been developed and new avenues of socializing among females have been created, yet the females are still expected to conduct their usual responsibilities around the farm and domestic responsibilities. Without the approval of husbands and other family members, the females cannot participate in Green Tourism activities, training sessions or seminars. Empowerment is happening to a degree; however, there are strong traditions that limit change.

In addition to limiting traditions, another major challenge seems to be that female farmers in rural areas are seniors above the age of 65, yet the Green Tourism's Female Empowerment Guidelines' main concerns are not specifically targeting female farmers in this age bracket. The Green Tourism initiatives instead support 'young female farmers who need help in child rearing responsibilities' and 'training and development of female farmers who are enthusiastic and capable' (MAFF 2007). Such policies and guidelines certainly help empower female farmers who are in their 30s and 40s; nevertheless, the reality is that ageing female farmers in rural areas will benefit little from such guidelines where traditional gender roles are still strong.

Although the case of Kunisaki Peninsula may not be generalized to many other agritourism projects, there are some findings that can also be seen in other communities. In a traditional rural society, females often bear the overwhelming responsibility for domestic affairs and work in the field. However, each community has different issues; for example, those in Kunisaki Peninsula are ageing farming communities. The existing government guidelines help very little to change the traditional mindset of older farmers in terms of the importance of gender equity within agricultural tourism. Even though Green Tourism was partly conceived to empower female farmers through the recognition of the work they have been doing, leading to financial independence and skill development, without assistance from male farmers, female farmers are trapped between existing responsibilities (agricultural production and domestic responsibilities) and the added responsibilities of Green Tourism activities. A society like Japan that relies on government direction and initiatives requires a well-structured public support system in which both the male and female members of society can see the benefit of these empowering projects.

References

Costa, J.A. (2005) 'Empowerment and exploitation: gendered production and consumption in rural Greece', *Consumption, Markets and Culture*, 8(3): 313–23.

Excite News (2008) Kihonteki nougyou-juujisha hitari ni hitori ga 65-sai ijou [one in two farmers are over 65 years old]. Economic News. Online. Online. Available at: www.excite.co.jp/News/politics/20080519/Economic_pol_080517_089_4.html (accessed 15 May 2008).

Gentry, K.M. (2007) 'Belizean women and tourism work opportunity or impediment'?, *Annals of Tourism Research*, 34(2): 477–96.

Honkawa (2008) Nougyousha no kourei-ka [ageing farmers], Honkawa Data Tribune, Shakai Jitsujou deta zuroku [Social facts in visuals]. Online. Available at: http://www2.ttcn.ne.jp/honkawa/0530.html (accessed 15 August 2008).

Kinnaird, V. and Hall, D. (eds) (1994) *Tourism: A Gendered Analysis*, Chichester: Wiley.

Kumamoto Prefecture (2003) Josei, Koreisha Muraokoshi Suishin Jigyou Jigyou Hyouka [Review of the Community Rejuvenation Project by Women and Elderly]. Online. Available at: www.pref.kumamoto.jp/invited/opinion/seisaku_hyoka2/h15/links/01/01-001-10.pdf (accessed 15 August 2008).

Ministry of Agriculture, Forestry and Fisheries (MAFF) (2007) Nourinsuisan-sho Danjo Kyodou Sankaku Suishin Honbu Dai 1-kai Kaigou (Heisei 19-nen 4-gatsu 9-ka Kaisai) [Ministry of Agriculture, Forestry and Fisheries Gender Equity Promotion Headquarter: The First Meeting on 9th of April. 2007]. Online. Available at: www.maff.go.jp/danjo/19honbu.htm (accessed 15 August 2008).

Momsen, J.H. (2004) *Gender and Development*, London: Routledge.

Miyagi, M. (2001) Chapter 9: Joseitachi no enpawamento [Female Empowerment] in Y. Iwasaki and M. Miyagi (eds) *Seikou-suru Nouson Josei Kigyou: Shigoto, Chiiki, Jibun-zukuri* [Successful Rural Female Entrepreneurships: Making of jobs, regions and self], Tokyo: Ie no Hikari Kyokai: pp. 209–18.

Oita Prefecture (2009) Life Expectancy Statistics. Online. Available at: http://www.pref.oita.jp/12000/toukei/jumyou.htm (accessed 21 April 2010).

Takahashi, I. (2002) *Koreisha to Chiiki Nogyo* [Ageing population and Regional Farming], Tokyo: Ie no Hikari Kyokai.

Telfer, D.J and Wall, G. (1996) Linkage between tourism and food production, *Annals of Tourism Research*, 23(3): 635–53.

Telfer, D.J. (2000) 'Agritourism – a path to community development? The case of Bangunkerto, Indonesia', in G. Richards and D. Hall (eds) *Tourism and Sustainable Community Development*, London: Routledge: pp. 242–57.

Telfer, D.J. and Sharpley, R. (2008) *Tourism and Development in the Developing World*, London: Routledge.

Torres, R. (2003) 'Linkages between tourism and agriculture in Mexico', *Annals of Tourism Research*, 30(3): 546–66.

Torres, R. and Momsen, J.H. (2004) 'Challenges and potential for linking tourism and agriculture to achieve pro-poor tourism objectives', *Progress in Development Studies*, 4(4): 294–318.

Part II

Building tourism and agriculture linkages: challenges and potential

6 Sustainability on a plate: linking agriculture and food in the Fiji Islands tourism industry

Tracy Berno

Introduction

One of the most important areas of the tourism industry is that of food production and consumption. Ironically, however, despite the fact that all tourists eat as part of their tourist experience, this is an area that has received little attention in the tourism literature. Much has been written about sustainable tourism, but few initiatives have been undertaken to consider the relationship between agriculture and sustainable tourism. Yet better linkages between agriculture and tourism not only can result in higher levels of economic retention but can also contribute significantly to the ethos of sustainable tourism.

Increasingly, as destinations seek to differentiate themselves in the market, a distinctive local cuisine can be used as a tool for promotion. This further serves to reinforce the increasing desire of tourists for 'authentic' experiences. Sustainable cuisine, which supports local agricultural production, can be an integral tool for sustainable tourism.

Food is an essential component of the tourism industry. It is obvious and it warrants mention that all tourists eat when they travel, and dining is consistently rated as one of the most popular tourist activities (National Restaurant Association 2003). For example, recent research in Australia found that dining at a restaurant/cafe was the top activity for international visitors to that country (Tourism Research Australia 2006).

Food and beverage consumption represents a significant part of tourist expenditure. It is estimated that 20 per cent of tourist expenditure in Fiji (FJ$ 150 million) is spent on food (Sustainable Tourism Development Consortium [STDC] 2007). Yet for many developing countries, including Fiji, food represents one of the highest areas of economic leakage in tourism. The degree to which tourism in a country relies on imported foods can significantly affect the social and economic impacts of tourism. Importing foods results in a loss of foreign exchange earnings and opportunities to expand and modernize local food production and processing. Potentially, this may result in a loss of local income and employment (Telfer and Wall 1996). In Fiji, the need to link local food to the tourism sector was identified as early as 1973 (Belt Collins 1973) and was highlighted in the 1976 Seventh Development Plan:

> Vegetable producers are keen to take advantage of the increasing local and tourist requirement ... Determined effort will be made to improve further the quality and supply of vegetables. Government will provide specialist extension advice on recommended varieties, fertilizers and chemicals.
>
> (Cited in Young and Vinning 2006: 39)

Despite an abundance of locally produced foods and food products, in most South Pacific Island nations, much of the food served in the tourism sector is imported. There are many reasons for the reliance on imported agricultural products in South Pacific Island countries. Availability, price, consistency, quality of local products, lack of technology, infrastructure and finance are some of the reasons cited (Berno 2003; Roa 2006; Salvioni 2007; Viet 2007; Young and Vinning 2006). Others have suggested that tourists demand foods that are often not grown in the host region or that tourists are reticent to try local foods or local cuisines (Berno 2003; Gomes 1997). Also posing threats to sustainability are the 'international' menus (which often include 'parodies' of local dishes rather than authentic local food) favoured by many large hotels and resorts and the proliferation of transnational fast food and restaurant chains at the expense of small locally owned enterprises (Parkinson 1998; Swarbrooke 1999; Torres 2003).

Despite these impediments, however, the benefits of tourism to a destination can be enhanced by expanding the backward economic linkages through increasing the amount of local foods used by the tourism industry (Telfer and Wall 1996). Creating and strengthening these linkages between tourism and food production sectors can provide a proximate market for locally produced products, while at the same time, enhancing tourists' experiences by providing them with the opportunity to consume high-quality local produce (Boyne *et al.* 2001). Enhancing linkages between agriculture and tourism presents significant opportunities for stimulating local production, retaining tourism earnings in the locale and improving the distribution of economic benefits of tourism to rural people (Torres 2003).

The benefits of increased linkages between tourism and agriculture go beyond just what is on the plate and include the generation of a range of both direct and indirect demands for agricultural products and services related to tourism (Fox and Cox 1993). The effects of these demands can result in a variety of positive outcomes, including a reduction in product miles and other waste, enrichment of localities and economic links, more attractive, vital and viable rural areas, a more vibrant and locally distinctive tourism and greater economic and social well-being for the host community.

The role of food and cuisine in sustainable tourism

Despite Berno's (2003) and Gomes' (1997) observations that some tourists are reluctant to try local foods and cuisines, an increasingly significant number of travellers are stating that food is a key aspect of the travel experience and that they believe experiencing a country's food is essential to understanding its culture (Condé Nast Publications, Inc. and Plog Research 2001, cited in Deneault 2002). In the wake of globalization and the 'homogenization' of tourism experiences, increasing numbers of tourists are seeking 'authentic' experiences. Food, or gastronomy, is one means of expressing authentic local culture or heritage (Haukeland and Jacobsen 2001; Richards 2002).

The increasing appeal of gastronomy and locally sourced cuisine in tourism results in part from processes in society at large. People's interests in food quality, ecological concerns about the needs for increased sustainable agricultural practices, health and nutrition concerns, a more sophisticated knowledge of food and beverages and acquired information about different cuisines are affecting tourists' expectations and behaviours when they travel (Haukeland and Jacobsen 2001).

The tourism industry can take advantage of the growing interest in sustainable food systems and sustainable cuisine by promoting and using more local products throughout the industry, while at the same time, meeting travellers' needs for an authentic, quality experience. By forging stronger linkages between agriculture and tourism through the development and promotion of sustainable cuisine, a symbiotic relationship between these sectors can be established.

The 'farm-to-fork' concept

By adopting the principles of sustainable cuisine, critical linkages between the tourism industry and local agricultural production can be forged, thus promoting further a holistic approach to sustainable tourism.

One way of operationalizing the articulation between the areas of sustainable agriculture, sustainable cuisine and the tourism industry is the development and promotion of the 'farm-to-fork' concept, which can support agricultural production by increasing demand for local products. The farm-to-fork concept promotes a high-quality tourism product through a value chain, which supports the use of local agricultural products within the tourism industry (Centre for Corporate Citizenship at Boston College [CCCBC] 2002). The farm-to-fork concept does not have to stop at just the dining experience, it can also lead to a range of related direct and indirect agritourism activities and products, including floriculture, food festivals, farm visits, factory tours and value added products such as souvenir food merchandise, thus further enhancing the benefits to the local community.

Like most initiatives to support sustainable development and to operationalize the farm-to-fork concept successfully, a broad range of stakeholders must be involved. These include interests such as relevant government ministries (e.g. tourism, agriculture, fisheries and environment), national food production, pastoral and agricultural associations, providers of tourism and hospitality education, chefs and/or hotel management, national and local tourism organizations, communities of interest (i.e. rural producers and individuals with relevant expertise) and others with interests in rural development, agricultural and/or tourism sectors. Implemented well, the farm-to-fork concept links local agricultural producers, sustainable cuisine and the tourism industry, resulting in positive outcomes for a broad range of beneficiaries.

The Caribbean shares many tourism attributes with the island nations of the South Pacific, including the need to enhance retention of the tourism dollar. Several farm-to-fork initiatives have been implemented in the Caribbean as a means of enhancing economic retention. The Farmers Programme, implemented and supported by the Sandals Group of all-inclusive resorts is an oft-cited example of a successful farm-to-fork initiative (see, e.g. McBain 2007; Meyer 2007).

The Sandals Group is a large all-inclusive resort chain with properties in several Caribbean countries. Sandals' Farmers Programme in Jamaica was implemented in 1996, with the aim of developing good working relationships between farmers and hotels by improving the quality and diversity of produce, developing formal pricing arrangements and improving communication linkages between farmers and hotels (Ashley *et al.* 2007; Rural Agricultural Development Authority [RADA], n.d.). The project started in 1996 with ten farmers supplying two hotels. By 2004, 80 farmers were involved. Based on the success of the Jamaica initiative, Sandals has replicated the programme in other Caribbean countries, including St. Lucia and Antigua (Ashley *et al.* 2007; McBain 2007) (see Chapter 8).

Agriculture, food and tourism – the context of Fiji

Despite the potential positive benefits to Fiji of a concept such as farm-to-fork, as discussed in preceding sections, increasing the linkages between tourism and agriculture has numerous challenges that must be overcome. There is no simple formula for increasing the use of local agricultural products in the tourism industry. As Torres (2003) points out, the examination of tourism and agriculture must be situated within the local farming development context of the host country destination.

There are numerous examples of successful strategic alliances between primary agricultural producers and the tourism industry internationally (see, e.g. Hui Mea'ai project in Kauai and Hawaii described in CCCBC 2002 as well as Ashley *et al.* 2005, 2007). Unfortunately, however, not all tourist destinations have the benefit of the technology, transport infrastructure or technical advice to support these projects, hence, the importance of considering the context. In many instances, such as in the small island nations of the South Pacific, including Fiji, there are challenges to the successful implementation and sustainability of these types of projects. Some Pacific Islands lack the basic natural resources for significant agricultural production. The very attributes that make them desirable tropical holiday destinations may mean that they are unable to support agricultural production. Small, low-lying coral atolls have little topsoil. That which is there is often of poor quality and high salinity, thus unable to sustain anything other than the most basic subsistence agriculture. In many, if not most of the Pacific Islands, there is a lack of transportation infrastructure, technology lags behind that of more developed countries and there is a lack of technical advice and communication between key stakeholders.

Fiji's economy relies primarily on tourism, agriculture (mainly sugar cane for export) and manufacturing. Like many other small island developing nations, Fiji has looked towards tourism as the main means for economic development, and tourism is now a critical mainstay of the local economy. Since 1989, tourism has generated more foreign earnings than any other sector, and tourist arrivals have increased steadily over the years. Fiji is now the main centre for tourism in the South Pacific region. The contribution to Fiji's employment is estimated at 20 per cent of the total employment figure for 2001 with around 40,000 jobs (direct and indirect employment) and a contribution to the GDP of about 19 per cent, which is more than any other single industry (Narayan and Prasad 2003).

Although Fiji Islands' socio-economic development status has improved significantly over the last three decades, significant disparities still exist in the economic development process and distribution of benefits between rural and urban areas, as well as between different regions, particularly in the outer islands. Given the cultural traditions and pristine natural environment, tourism offers good development potential for outer islands and rural areas. Despite these attributes in the rural areas and outer islands, the majority of tourist activities and expenditure (87 per cent) is concentrated in the southern and western sides of the main island of Viti Levu (Ministry of Tourism 2003). As such, the economic benefits of tourism are also concentrated in these high tourist density areas. Many of the poorest areas are those with the least number of tourist arrivals and hence, the lowest tourist expenditure.

Several of the efforts undertaken in Fiji to maximize the economic benefits derived from tourism have concentrated on increasing the number of tourist arrivals, the average length of stay and the overall expenditure of tourists. Despite helping to boost the overall GDP of Fiji, these sorts of initiatives have had few salient benefits for the outlying (rural) regions. One potential way of enhancing the economic benefits of tourism to a destination is to expand the backward economic linkages by increasing the amount of local agricultural products used in the tourism industry.

Fiji is presented as a case study to give an example of some of the situational factors and issues that need to be addressed in promoting the articulation between local agricultural production, sustainable cuisine and the tourism industry in Fiji.

Linking agriculture, food and tourism in Fiji

Traditionally, tourism to the South Pacific has been predicated on the 'sun, sea, sand' image associated with larger scale mass tourism. In the South Pacific, this form of tourism has often resulted in a high degree of economic leakage and poor multiplier effect. One area of tourism in the South Pacific which has experienced a high degree of economic leakage is that of the food and catering sector. Like many small island developing nations, Fiji experiences a high degree of leakage of the tourism dollar (the most recent estimate is approximately 60 per cent) and much of that is attributed to leakage through imported foodstuffs in the hotel and resort sector. There is a lack of national-level data about how much food product is imported into Fiji to support the tourism industry. However, the Ministry of Tourism in Fiji suggested that up to 80 per cent of food in the tourism industry is imported, stating that, 'There is a serious problem' (Ministry of Tourism 2004a: 3). In dollar terms, it has been reported that approximately FJ$ 30 million is spent by the tourism industry in Fiji for importing products from overseas, many of which could be sourced and supplied locally (Hildebrand 2003). One major resort purchasing officer reported that his resort imported up to 40 per cent of their fruits and vegetables from overseas (personal communication 2003). Although some amounts of local products were found to be used in the tourism sector, research suggested that this use was heavily skewed towards three main areas: fruits/vegetables for breakfast and/or inclusion in 'western menus' (e.g. salads and vegetables to accompany meals), traditional Fijian *lovo* (earth oven) nights and staff meals (Fiji Hotel Association 1999).

The need to increase the linkages between tourism and agricultural production to address the retention of tourism revenue has been identified by the Ministry of Tourism (2002, 2003), by the South Pacific Tourism Organisation (SPTO, 2003), and in the Pacific 2020 project (McGregor 2006). The *Fiji Tourism Development Plan 2007–2016* also highlights the need to create and sustain these linkages (STDC 2007). This has been reinforced further through recent initiatives such as the Reserve Bank of Fiji's announcement of the introduction of policies to link borrowing from commercial banks and other lending institutions by hotel operators to local food production and environmental protection. The Reserve Bank will further support this policy by giving an annual Prime Minister's award to the most successful and innovative hotel to achieve high local value added and being environmentally friendly. The Reserve Bank will also give a 6-month scholarship to the most promising graduate from the tourism industry schools to study local food processing (Reserve Bank of Fiji 2009).

From an economic point of view, there is an urgent need for the tourism industry in Fiji to optimize the use of locally produced food. Despite this need, however, there is a range of impediments that inhibit significantly increased use of local products in the tourism sector.

A great idea – but will it work in Fiji?

Fiji is a small, developing island nation, comprising over 300 islands, with a population of just over 837,000 (Fiji Islands Bureau of Statistics 2007a). The majority of the population comprises indigenous Fijians and Indo-Fijians (the descendants of indentured labourers brought to Fiji in the late 1880s to work in the sugar cane fields). There are also small minorities of Chinese, Europeans and other Pacific Islanders in Fiji.

Traditional indigenous ways of life and community structures are still very strong and influential in Fiji. In total, 86 per cent of the land in Fiji is under traditional indigenous Fijian ownership, which means that developers cannot buy native land outright; rather, they must negotiate leases and permission to develop. Through the leases, restrictions are placed on the type, scale and distribution of the benefits of development initiatives. This has a range of impacts on both tourism and agricultural development in the country. Land tenure–related disputes in Fiji over the last decade have negatively affected agriculture (Lal and Reddy 2002), tourism (Waqaisavou 2002) and forestry (King 2001), with these sectors having been frustrated by impediments and losses due to disputes (Murti and Boydell 2008). Community-based initiatives in ecotourism, marine life management and agriculture have also been adversely affected in the past because of unresolved conflicts within landowning communities (Warner 2000). Murti and Boydell (2008) have suggested that these Fijian landownership conflicts, often based on confusion over property rights and related issues, occur within resource owning communities, between communities and external parties and among external parties. These conflicts stem from differing views on ownership, land tenure and property rights and have led to delayed implementation of critical environmental management plans, loss of economic benefits and disintegration within landowning units.

The tourism industry now attracts just over 500,000 visitors to Fiji per annum (Fiji Islands Bureau of Statistics 2009), most of whom stay in resorts in the drier west of the main island of Viti Levu, which is one of the main areas for agricultural production. Although many of the small islands in Fiji are unable to support significant agricultural production, Viti Levu, the main centre for tourism, is a large fertile island of multiple climates and is able to support a broad range of agricultural initiatives.

As a result of the diverse composition of the population, there are a number of cuisines in Fiji, many of which have been adopted and adapted from other cultures. Indo-Fijian, Chinese and indigenous Fijian cuisines are all available, as are foods from many Western and Asian traditions.

Most of these foods are represented in some form on the menus of the hotels and resorts, though for the most part, menus reflect a bias towards 'Western' style foods, which often require imported ingredients. Despite the potential appeal to the culinary tourist, or the tourist seeking a more authentic food experience, local Fijian ingredients such as ota (wild river fern), duruka (heart of local sugar cane), breadfruit and jackfruit are rarely seen on tourist menus. The Pacific Islands, including Fiji, however, are not generally regarded for their cuisine, and it would be fair to say that they are not considered a destination for culinary tourism. At its most unkind, the food of the South Pacific has been described as '… lousy. You can't eat it' (Ochef[1]). When one thinks of Fiji, it is inevitably the ubiquitous 'sun, sea, sand' image that comes to mind rather than culinary experiences.

As the Ministry of Tourism has bluntly indicated, 'Food is an ongoing problem' (Ministry of Tourism 2004b: 80) [and] '… remains an Achilles Heel in the Fijian market' (Ministry of Tourism and Stollznow Research 2005: 63). Only an approximate third of the visitors to Fiji gave the food a 'very good' rating, and only 4 per cent of visitors indicated that 'good food' was one of their positive impressions of Fiji. Research by Cummins (2002: 6) found that tourists interviewed said 'that they had tried the local foods [at hotel *lovo* nights or at street cafes] but did not wish to try them again'. As pointed out by the Ministry, 'It is felt very strongly that work must … start on better understanding the food requirements of visitors to Fiji' (Ministry of Tourism 2004b: 80).

Despite these negative assertions, there are opportunities to increase the linkages between tourism and agriculture through the development of a distinctive sustainable cuisine, a 'Fiji/Pacific cuisine' (see, e.g. Oliver *et al.* 2010), as a means of operationalizing the farm-to-fork ethos. Stakeholder consultation with farmers, chefs, tourism operators and relevant government ministries included a SWOT (Strengths, Weaknesses, Opportunities, Threats) analysis of agriculture – tourism linkages. Through the SWOT analysis, a range of facilitators and barriers to forging articulations between sustainable agriculture, sustainable cuisine and the tourism industry in Fiji were identified (Berno 2003, 2006). These are summarized in Table 6.1.

Stakeholder consultation has suggested that there is a good deal of interest in and support for increasing the use of local ingredients in the Fiji tourism industry. Key stakeholders such as the Ministry of Tourism, the Fiji Island Hotel Association, the SPTO, growers and suppliers, vocational training institutions and tourism operators themselves have expressed a strong desire to see an increase in local products (Berno 2003, 2006), particularly if they are used in innovative ways that are appealing to the tourist palate. Fiji is a fertile country that has the potential to operationalize the farm-to-fork concept in the tourism sector. Recent changes in agricultural land, including the non-renewal of agricultural leases (mainly cane farms), mean that there is an availability of potential land and labour to support such initiatives. There is also a great need for poverty alleviation and sustainable livelihood projects in rural and outer island areas, with several non-governmental organizations interested in rendering assistance for these initiatives.

Despite these opportunities in Fiji, there are a several significant barriers to further development towards increasing the linkages between the agricultural and tourism sectors. These barriers have contributed to the lack of sustainability and significant progress in increasing the use of local products in the tourism sector. In a previous attempt in 1986 to address the situation, the then Tourism Council of the South Pacific (now the SPTO) commissioned the writing of a tropical foods cookbook for hotels, *Cooking the South Pacific Way – A Professional Guide to Fiji Produce*. To ensure supply of local products to support the recipes in the book, the project was linked with primary agricultural production, with an expectation that the Ministry of Agriculture would assist through encouraging the increased production and supply of the foods discussed in the book. The projected outcome of this effort was for hotels to be able to obtain adequate local products to help reduce the reliance on imported fruits and vegetables.

In total, 1,500 copies of the book were produced and distributed throughout the region, mostly in Fiji. Initial feedback was positive with the recipes appearing to be useful to chefs, and most hotels began to use some locally grown vegetables on a more regular basis. However, many chefs complained that the supply of local

Table 6.1 Facilitators and barriers to the implementation of the farm-to-fork concept in Fiji (Berno 2004, 2006)

Facilitators		*Barriers*	
Strengths	*Opportunities*	*Weaknesses*	*Threats*
• The Fiji context offers potential for improving the variety, quality and value of the food experience in Fiji • An abundance of well-trained chefs working in the industry • Increasing interests in organic production in the agricultural sector • Strong linkages between cuisine and culture in Fiji • Food is a year-round, 24-hour-a-day product • Almost unexploited in terms of a tourism attribute in Fiji • Broad range of good local products (food and floriculture) available • Proximity of the majority of the tourism industry to the main agricultural growing regions	• Work with stakeholders to develop and implement fully the Fiji-Grown concept • Promote sustainable cuisine to tourists by working with industry and producers to create local and export-ready products • Increase Fiji Grow promotional activities and partnerships with the industry (major hotels, destination marketing organizations, transport and accommodation sectors) • Creation of 'value-added' products such as souvenir food items, farm tours and so on • Capitalize on the excellent growing conditions in Fiji to grow both indigenous products and high-demand exotics • Changes in the agricultural sector (i.e. cane farm leases)	• Poor perception of 'Fiji/Pacific cuisine' internationally (minimal consumer demand) • Lack of a clearly defined 'Fiji/Pacific cuisine' • Many local treatments of food products are unacceptable to the tourist palate • A lack of interest/confidence on the part of some chefs • Concern on behalf of growers about sustainability of demand • Lack of resources for education and training on the preparation of indigenous products for the tourism sector • Lack of organizations focusing on the development of local products for the Fiji tourism product • Poor communication throughout the supply chain • Uncooperative 'gatekeepers' • Poor infrastructure • Lack of private/public sector communication/alliances • Poor access to farming technologies as well as IT	• Lack of sustainability of previous initiatives • Tourist preferences for familiar (i.e. non-Pacific) foods • Irregular availability and poor quality control in the production, processing and distribution of local food products for the tourism sector • Budget constraints making it difficult for some operators to adopt Fiji Grow (e.g. use of fresh flowers and fruit baskets in hotel rooms, flower garlands upon arrival etc.) • Negative perception of food in Fiji may make the concept of Fiji Grow challenging to promote • Threats to sustainability (fish blasting, overfishing, and effect on subsistence) • Land lease issues

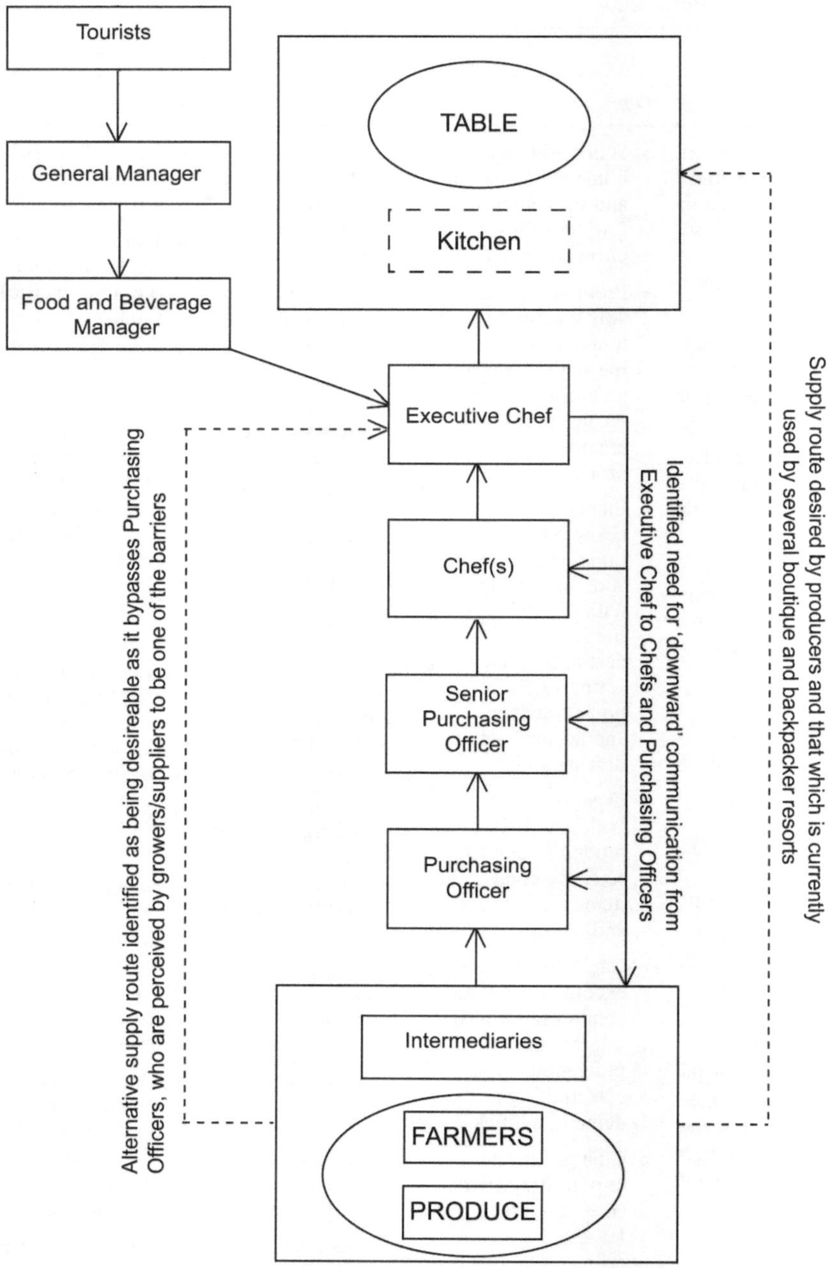

Figure 6.1 Facilitators and barriers identified in the supply chain.

fruits and vegetables was irregular and of inconsistent quality. Unfortunately, essential linkages between hotel chefs, purchasing officers and the Ministry of Agriculture lacked coordination, and the project, despite its initial encouraging success, was not sustainable.

The situation today remains much the same. However, it is worthwhile noting that there are projects that have met with some success (discussed further in following paragraphs). Most of these projects have tended to circumvent the lengthy supply chain within which many of the larger resorts and hotels are required to work (Figure 6.1). Within these supply chains, Purchasing Officers in particular were identified as barriers to increasing use of local products, with one supplier describing them as the 'barracudas between supplier and chef'.

In addition to the poor communication and weak linkages between key positions in the supply chain, there is also a lack of requisite infrastructure to support distribution of produce to the hotels and resorts. In addition to this, few of the rural producers have access to IT to support effective communication between government, the suppliers and the hotels. Milne and Mason (2000) identified use of IT as a means of enabling linkages between these sectors.

Although producers and suppliers were overwhelmingly in support of initiatives to support the increased use of their products in the tourism sector, they too expressed some concerns. One grower expressed concern over the dynamic nature of hotel menus, stating that menu changes would mean that he would be left with surplus produce. Similarly, another grower expressed concern about the length of time that it took to get new items on the menu, suggesting that it could be difficult to respond to restaurant needs. It was reported that on average, it took about 6 weeks to get a new item on a hotel menu. By that time, a particular product may no longer be at its peak, or its season may have passed, so the opportunity to supply it would be lost.

Finally, it is also important to note that despite all the positive benefits of increasing the use of local products in the tourism industry, this laudable outcome can potentially pose threats to sustainability/sustainable tourism. For example, increased demand for local seafoods could place pressure on existing fish stocks. This is particularly relevant for high tourist density areas in Fiji, which have already experienced environmental degradation and loss of marine life due to tourism development (personal communication 2003). Anecdotal evidence exists to suggest that some Fijian villagers are using the highly environmentally damaging method of dynamiting for fish extraction, as they perceive that it is the only way that they can keep pace with demand for fresh fish for nearby resorts (personal communication 2004). There is also concern that an increase in demand for local agricultural products could upset the traditional subsistence lifestyle of suppliers, thus posing threats to rural food security (personal communication 2003). As the key point of the farm-to-fork proposal is to address the linkages between sustainable agriculture and sustainable tourism, any initiative must be designed to negate or mitigate the potentially deleterious effects such as those discussed in preceding paragraphs.

Making the links

It is clear that the potential exists in Fiji to forge stronger relationships between the agriculture and tourism industries. Research by Berno (2005) suggested that there are examples of tourism–agriculture linkages in Fiji that have had varying degrees of success. Seven different mechanisms that supported agriculture–tourism linkages were identified:

- Market – tourism operators buying produce directly from local markets
- Product-led approach – hotels establishing relationships or contracts with individual suppliers often for specific products
- Surplus – market stallholders approaching hotel operators with surplus produce for sale
- Cooperative – the formation of farmers' cooperatives to supply hotel operators
- Creation of demand – the introduction of menu items using local indigenous products
- 'Grow your own' – the use of on-site hotel gardens to supply the accommodation with a limited range/amount of fresh produce
- 'Boutique' – organic and/or hydroponic gardens attached to high-end accommodation
- Strategic business unit model

For example, some of the 'top-end' boutique resorts in the country have developed distinctive cuisines, and several grow their own, often organic, produce to support their requirements. These resorts have excellent international reputations for the quality of their foods. However, owing to their exclusivity and usually remote geographic location, the impacts of their initiatives on the Fiji economy, the local community and the tourism industry as a whole are relatively minor.

On the other end of the tourism scale, budget 'backpacker'-type accommodations often source many of their food products locally, as this is often a less expensive option than purchasing costly imported goods. Additionally, many backpacker operations in Fiji are locally owned and managed. As such, staff often do not have the formal training or professional skills to use exotic ingredients and therefore rely on local products cooked in more traditional ways. Backpackers comprise 13 per cent of the tourism market in Fiji (Ministry of Tourism and Stollznow Research 2005) and in a recent survey, 70 per cent indicated that 'experiencing Fijian culture' was an important factor in their choice to travel as a backpacker/independent traveller (Jarvis and Hobman 2006). This presents a promising area for further development of food as part of the cultural experience for this tourist group. A recent study by Viet (2007) reinforced the potential of the backpacker market for purchasing local products. Viet found that more than 60 per cent of backpacker hoteliers ranked the price of produce as the most important criterion for purchasing decisions. As such, backpacker operations have potential as a niche market for produce that cannot be sold to the larger and more expensive hotels due to restrictions in quality.

The bulk of inbound tourism to Fiji can be described as 'mass tourism'. This is often the most difficult sector of the tourism industry in which to make inroads in terms of introducing farm-to-fork linkages. Even in this area, however, some producers have had success in increasing supplies to the larger resorts. At the time that the research was undertaken, one initiative, a farming project spearheaded by a multinational resort chain, was successful in establishing a structure to facilitate such linkages (Naivalumaira 2005). This model, which applied a strategic planned approach to the establishment of a sustainable commercial production and supply model, is one that could provide a template for further developments of this nature. However, the requirements for the ongoing sustainability of such initiatives need to be identified.

Other initiatives in Fiji included the formation of a local agricultural cooperative of farmers on the main island of Viti Levu. This approach has also been introduced in Jamaica where six cooperatives were set up under the Sandals/RADA initiative and more than a decade later five are still functioning (Rhiney 2008). The cooperative had the capacity to supply fresh produce directly to some of the larger resorts in the high tourist density areas. This cooperative achieved some success due to (1) its close proximity to the main tourism region in the country, (2) good communication between all stakeholders in the supply chain and (3) a collective approach. The Executive Chef of one of the larger chains of hotels spearheaded another initiative that led to the highlighting of more local ingredients on the menus throughout the properties in the group. This initiative was successful owing to the personal interest of one of the key gatekeepers in increasing the use of local ingredients in innovative ways and the redesign of the menus to reflect this. The chef's goals were further reinforced through other industry activities, such as a 'local ingredients' class in the Fiji Salon Culinaire. Unfortunately, however, the eventual departure of this Executive Chef saw a change in the menu and a return to more traditional 'Western' fare. Other similar initiatives introducing more use of local ingredients as part of a contemporary Pacific cuisine have also not been sustained.

On a larger scale, there are numerous salient benefits to a broad range of stakeholders should farm-to-fork initiatives incorporating a Pacific sustainable cuisine be successfully developed and implemented industrywide. Such initiatives could contribute to sustainable rural livelihoods and poverty alleviation, support the agricultural sector and reduce economic leakage by addressing the ongoing need to increase the production and use of local products in the main 'mass' tourism sector through

- developing sufficient, reliable, quality local foods for hotels (both indigenous and exotic products);
- training chefs in the cooking and presentation of local produce;
- providing chefs and tourism/hospitality educators with education and training about local food recipes suitable for modern hotel menus and

- using local quality cuisine as a marketing tool for hotels adopting the program, while
- ensuring that the health and nutrition of rural producers is not compromised by the diversion of food supplies to the hotels (Berno 2003).

However, lessons learnt from past efforts as well as the situational analysis suggest that the way forward requires careful consideration of the particularities of the Fiji context. The successful linkages already achieved can serve as examples as to how these linkages can be implemented and sustained. Further work and research, however, are required to take the initiative countrywide.

Note

1 On-line discussion on South Pacific foods in the Ochef site: www.ochef.com/95.htm.

References

Ashley, C., DeBrine, P., Lehr, A. and Wilde, H. (2007) *The role of the tourism sector in expanding economic opportunity*, Boston: The Fellows of Harvard College, Overseas Development Institute, International Business Leaders Forum.

Ashley, C., Gareth, H. Poultney, C. McNab, D. and Harris, A. (2005) *How to…? Brief 1: Boosting procurement from local businesses*, UK: Overseas Development Institute and Business Linkages in Tourism.

Belt Collins (1973) *Tourist development program for Fiji*, Boulder, CO: Belt Collins and Associates.

Berno, T. (2003) 'Fiji Grown: From farm to restaurant', keynote address presented at the FAO/Ministry of Agriculture Fiji Grown workshop, 26–28 February, Korolevu, Fiji.

Berno, T. (2004) 'New approaches to sustainable tourism: Optimising the linkages between agriculture and tourism in the Fiji Islands', paper presented at the Association of American Geographers conference, 17–22 March 2004, Philadelphia.

Berno, T. (2005) 'Not because of the food – Can Pacific cuisine become an attribute of the Pacific tourism product?', paper presented at the Council of Australian Universities in Tourism and Hospitality Education Conference, 2–5 February, 2005, Alice Springs, Australia.

Berno, T. (2006) 'Bridging sustainable agriculture and sustainable tourism to enhance sustainability', in G. Mudacumura, M.S. Haque and D. Mebratu (eds) *Sustainable development policy and administration*, New York: Taylor & Francis.

Boyne, S., Williams, F. and Hall, D. (2001) 'Rural tourism and food production: Opportunities for sustainable development', paper presented at the RICS Foundation Roots 2001 Conference, 12–13 November, London.

Centre for Corporate Citizenship at Boston College (CCCBC) (2002) *A Productive Partnership: The Kauai Marriot Resort and Beach Club and the Kauai Food Bank*, Boston: The Centre for Corporate Citizenship.

Cummins, K. (2002) *Agriculture, food and nutrition in the developing world*, unpublished manuscript.

Deneault, M. (2002) *Acquiring a taste for cuisine tourism: A product development strategy*, Ottawa: Canadian Tourism Commission.

Fiji Hotel Association (1999) *Report on a survey to determine the kinds and amounts of locally produced foods used by Fiji hotels and resorts*, Unpublished report.

Fiji Islands Bureau of Statistics (2007a) '*2007 Census of population and housing*', Suva, Fiji: Fiji Islands Bureau of Statistics.

Fiji Islands Bureau of Statistics (2009) 'Summary of visitor arrivals and departures 2007'. Online. Available at: www.statsfiji.gov.fj/Tourism/Visitor_Arrivals.htm (accessed 14 May 2009).

Fox, M. and Cox, L.J. (1993) 'Linkages between agriculture and tourism', in M. Khan, M. Olsen and T. Var (eds) *VNR's Encyclopaedia of Hospitality and Tourism*, New York: Van Nostrand Reinhold, pp. 910–17.

Gomes, A.J. (1997) 'Integrating tourism and agricultural development', in France, L. (ed.) *The Earthscan Reader in Sustainable Tourism*, New York: Earthscan Publications, pp. 187–95.

Haukeland, J.V. and Jacobsen, J.K.S. (2001) 'Gastronomy in the periphery: Food and cuisine as tourism attractions at the top of Europe', paper presented at the 10th Nordic Tourism Research Conference, Vasa, Finland, 18–20 October 2001.

Hildebrand, S. (2003) 'Agriculture to have new heyday', *Fiji Times*, Wednesday 2 April, p. 22.

Jarvis, J. and Hobman, E. (2006) *Flashpackers – backpackers and independent travellers in Fiji 2006*, Melbourne: Monash University, Tourism Research Unit and National Centre for Australian Studies.

King, T. (2001) *Global forest fire assessment*, Rome: Food and Agriculture Organisation of the United Nations. Online. Available at: www.fao.org/docrep/006/ad653e/ad653e00.htm (accessed 29 March 2010).

Lal, P. and Reddy, M. (2002) 'Proposed sugar industry model and landlease issue – will it be a case of old wine in new bottle?', paper presented at the South Pacific Land Tenure Symposium, Suva, Fiji Islands. Online. Available at: www.sidsnet.org/pacific/usp/landmgmt/SYMPOSIUM (accessed 29 March 2010).

McBain, H. (2007) *Caribbean tourism and agriculture: Linking to enhance development and competitiveness*, ECLAC Studies and Perspectives Series, The Caribbean, No. 2. New York: United Nations Publications.

McGregor, A. (2006) *Pacific 2020 background paper: Agriculture*, Canberra: Commonwealth of Australia.

Meyer, D. (2007) 'Pro-poor tourism: From leakage to linkages. A conceptual framework for creating linkages between the accommodation sector and "poor" neighbouring communities', *Current Issues in Tourism*, 10(6): 558–83.

Milne, S. and Mason, D. (2000) 'Linking tourism and agriculture: Innovative IT solutions', paper presented to the Regional Agro Tourism Conference, Tobago, April 2000.

Ministry of Tourism (2002) 'Fiji: The contribution of the tourism sector to the economy', in United Nations Economic and Social Commission for Asia and the Pacific, Committee on Transport, Communications, Tourism and Infrastructure Development, Fourth Session, 13-15 November, Bangkok, Bangkok: UNESCAP CTCTID, pp. 12–19.

Ministry of Tourism (2003) '*Draft development and marketing unit workplan 2003*', Unpublished manuscript, Suva, Fiji: Ministry of Tourism.

Ministry of Tourism (2004a) *Report on the food policy for the travel and hospitality conference, November 17th 2004*, Suva, Fiji: Ministry of Tourism, Planning and Research Section.

Ministry of Tourism (2004b) *International visitors survey*, Suva, Fiji: Ministry of Tourism.

Ministry of Tourism and Stollznow Research (2005) *International visitors survey*, Suva, Fiji: Ministry of Tourism.

Murti, R. and Boydell, S. (2008) 'Land, conflict and community forestry in Fiji', *Management of Environmental Quality: An International Journal*, 19(1): 6–19.

Naivalumaira, N. (2005) 'The land, its people and a vegetable patch', *Vanua*, 3: 5–6.

Narayan, P.K. and Prasad, B.C. (2003) 'Fiji's sugar, tourism and garment industries: A survey of performance, problems and potentials', *Fijian studies: A journal of contemporary Fiji*, 1(1): 4–27.

National Restaurant Association (2003) 'Travel and Tourism Facts'. Online. Available at: www.restaurant.org/tourism/facts.cfm (accessed 14 May 2009).

Oliver, R., Berno, T. and Ram, S. (2010) *Me'a Kai: The food and flavours of the South Pacific*. Auckland: Random House.

Parkinson, S. (1998) *Local food production committee*, unpublished manuscript, Suva, Fiji.

Reserve Bank of Fiji (2009) *Reserve Bank announces foreign exchange preservation policies*, Suva, Fiji: Reserve Bank Press Release, 25 April 2009. Online. Available at: www.reservebank.gov.fj/docs/Press%20Release%20No%2019%20RBF%20Announces%20Foreign%20Exchange%20Preservation%20Policies.pdf (accessed 14 May 2009).

Rhiney, K. (2008) 'Towards a new model for improved tourism-agriculture linkages? The case of two farmers' co-operatives in Jamaica', *Caribbean Geography*, 15(2): 142–59.

Richards, G. (2002) 'Gastronomy: An essential ingredient in tourism production and consumption?', in A. Hjalager and G. Richards (eds) *Tourism and gastronomy*, London: Routeledge, pp. 3–20.

Roa, M. (2006) *Challenges and issues in pro-poor tourism in South Pacific Island Countries: The case of the Fiji Islands*, Working paper No. 2006/27 (ELMS-8), Suva, Fiji: ELMS, School of Economics, University of the South Pacific.

Rural Agricultural Development Authority (RADA) (n.d.). Sandals/Farmers programme. Online. Available at: www.rada.gov.jm/site_information.php?ID=27 (accessed 29 March 2010).

Salvioni, C. (2007) How to make money by feeding the tourists: The case of Fiji, paper presented at the 105th EAAE Seminar, Bologna, Italy, March 8–10, 2007.

South Pacific Tourism Organisation (SPTO) (2003) *Regional tourism strategy for the South and Central Pacific. Part One: Strategy for growth*, Suva, Fiji: SPTO.

Sustainable Tourism Development Consortium (STDC) (2007) *Fiji tourism development plan 2007-2016*, Suva, Fiji: STDC.

Swarbrooke, J. (1999) *Sustainable tourism management*, London: CABI.

Telfer, D.J. and Wall, G. (1996) 'Linkages between tourism and food production', *Annals of Tourism Research*, 23(3): 635–53.

Torres, R. (2003) 'Linkages between tourism and agriculture in Mexico', *Annals of Tourism Research*, 30(3): 516–66.

Tourism Research Australia (2006) *Top ten: Activities. International market tourism facts, December 2006*. Online. Available at: www.reservebank.gov.fj/docs/Press%20Release%20No%2019%20RBF%20Announces%20Foreign%20Exchange%20Preservation%20Policies.pdf (accessed 12 May 2009).

Viet, R. (2007) *Tourism, food imports, and the potential of import-substitution policies in Fiji*, Unpublished report prepared for the Ministry of Agriculture, Suva, Fiji: Fisheries & Forestry.

Waqaisavou, T. (2002) 'Tourism on native land in Fiji', paper presented at the South Pacific Land Tenure Conflict Symposium, Suva, Fiji Islands. Online. Available at: www.sidsnet.org/pacific/usp/landmgmt/SMPOSIUM (accessed 29 March 2010).

Warner, M. (2000) *Conflict management in community-based natural resource projects: Experience from Fiji and Papua New Guinea*, London: Overseas Development Institute. Online. Available at: www.odi.org.uk/publications/working_papers/wp135.pdf (accessed 29 March 2010).

Young, J. and Vinning, G. (2006) *Fiji: Commodity chain study*, GTFS/RAS/198/ITA, Suva, Fiji: FAO.

7 Cracks in the pavement: conventional constraints and contemporary solutions for linking agriculture and tourism in the Caribbean

Benjamin F. Timms and Stern Neill

Introduction: agriculture, tourism and linked development

Caribbean development theorists have long advocated economic diversification and the creation of linkages between sectors of the economy to address the vulnerability of export dependence and to shore up foreign exchange leakages (Beckford 1972; Karagiannis 2004; Lewis 1955). However, for the most part, this has not occurred, with resultant disarticulated island economies, rising food import bills and a growing reliance on the tourism industry (Brohman 1996). Yet tourism provides immense potential to overcome these economic limitations through linkages with other sectors of the economy to address the vulnerabilities of plantation economies.

One such linkage, between tourism and agriculture in the small island developing states of the Caribbean, has been well articulated over the past 40 years, whereby increased demand from tourism should be met by increased supply from agriculture (Momsen 1998). Unfortunately, the empirical results of achieving these benefits, for the most part, have been less than stellar. The discrepancy between theory and reality has traditionally been blamed at the local scale on lack of supply from domestic agriculture, inconsistent demand from the tourist sector and, to a lesser extent, the weakness of marketing (Bryden 1973; Belisle 1983; Gomes 1993; Torres 2003; Timms 2006). Here, the theoretical scale of analysis is expanded to include global economic processes as significant contributing factors in explaining these local constraints in the Caribbean.

Although structural forces have contributed to the degradation of domestic agricultural production and increased foreign influence in the tourism industry, they can also periodically offer opportunities to proactive actors (Timms 2009). In this chapter, the current food crisis resulting from global agricultural restructuring is argued to create conditions conducive to linkage formation by providing an incentive for increased domestic agricultural production and more locally oriented tourist demand. However, such an opportunity may, once again, be squandered if marketing is not more fully understood and developed to facilitate the linkage between agriculture and tourism. This chapter concludes with the introduction of a framework for linkages we believe can serve as a way forward for future research and action that is based on the four 'Ps' of marketing: product, place, promotion and price.

Conventional constraints

Insufficient supply

The lack of supply from domestic agriculture is a key constraint to the formation of linkages with tourism in the Caribbean. Simply stated, domestic agricultural production is unable to satisfy the large quantities, expected quality and price points demanded from hotels and restaurants serving the tourist market (Belisle 1983; Gomes 1993). Small Caribbean producers have difficulty competing with large industrialized agricultural producers in the United States and European Union due to differences in the economies of scale, access to credit and use of technology (Timms 2009). It has also been argued that labour competition with tourism raises wages and, hence, production costs, contributing to the high price of local products (Dwyer and Forsyth 1993; Torres 2003).

The problems of domestic agriculture in the Caribbean began early in the colonial era, when the vast majority of agricultural production was structured for export with domestic food demand satisfied by imports (Williams 1970). During the 1970s, this plantation legacy was challenged through efforts to increase domestic agricultural production to stem the foreign exchange leakage created by unfavourable terms of trade between exports and food imports (Beckford 1972). Small farmers oriented towards the domestic market were supported by government efforts to increase access to land and capital for productive investment, import restrictions and tariffs to protect domestic production and extend services to diffuse technical knowledge (Best 1976; Witter 2004), with reported success in countries such as Jamaica (Hope 1986; Manley 1982).

Unfortunately, the focus on domestic production was short lived, as rising oil prices in the 1970s and a subsequent drop in prices for export commodities in the 1980s led to fiscal crises for many Caribbean states (Harker 1989). As a result, international lending agencies placed conditions on loans aimed to create fiscal austerity by cutting spending and promoting neoliberal economic development through international trade (Weis 2004). Budget cuts for agricultural support services led to the scaling back of extension services, agricultural marketing boards and supportive subsidies (Deep Ford and Rawlins 2007). International lending agencies, concerned about inflation, even required local banks to raise interest rates to such a degree (20–40 per cent) that farmers were unable to access credit for productive investment (Stiglitz 2003).

Furthermore, liberalized trade policies eroded import restrictions and tariffs, opening up domestic markets to subsidized imports, and currency devaluations were adopted to make Caribbean exports more competitive on the international market. Ironically, cutting import tariffs left governments with less revenue from import taxes and currency devaluation increased the costs for imported agricultural inputs, raising production costs and negating the competitive advantage this should have created by making imported food products more expensive (Deep Ford and Rawlins 2007; Witter 2004). Neoliberal economic restructuring intensified the challenges that domestic agriculture faced and contributed to supply-side problems for linkage formation with tourism and other sectors of the economy, with resultant foreign exchange leakages and rising food trade imbalances.

Inconsistent demand

With estimates of up to a third of tourist expenditure going directly to food production, the potential demand is indeed significant (Belisle 1983; Hall and Sharples 2003; Torres 2002). Demand is initially determined by the preference of tourists, with different types of tourists having varying demands for local agricultural products (Torres 2002). For example, tourists with longer visits are more likely to try local foods than tourists on shorter visits, who tend to prefer familiar imported food items (Organization of American States 1984). Additionally, hotels tailored towards wealthier tourists tend to pay more for presumed higher quality imports, bypassing local producers (Belisle 1984; Torres and Skillicorn 2004). However, the mass tourism package resorts aimed at more budget-conscious travellers seek out cheaper products, which have traditionally been imported as well. The demand for high quality or low prices creates a vacuum for local producers. A compounding element is the demonstration effect, whereby tourist preferences for imported products are adopted by local populations, contributing to an overall lack of demand for domestic products (Bryden 1973; Fisher 2004; McElroy 1986; Momsen 1998).

In addition to the demands of the tourists is the preference of the chef and/or food procurement officer. Torres' (2003) study in Quintana Roo, Mexico, found that foreign chefs purchased more imported food than local chefs, particularly due to the convenience of packaged foods and a lack of familiarity and comfort in using local products (Belisle 1984; see Chapter 8). Chefs also have concerns about food safety of local sources, whether or not these concerns are supported by evidence (Torres and Skillicorn 2004). Much of this is due to the unfamiliarity of chefs and tourists with local products and producers, which is in part explained by the social differentiation between elite tourists/hotel professionals and what are perceived to be traditional local farmers (Gomes 1993).

Contributing to the satisfaction of foreign preferences is the increase in foreign influence that trade liberalization has allowed. With liberalized trade policies, the availability of imported products has increased, both in terms of price and quality. Furthermore, these policies have created a concurrent rise in foreign investment and ownership of tourist establishments. Combined with the aforementioned preferences of foreign chefs, large foreign-owned hotels tend to rely more heavily on imported products than small locally owned hotels (Momsen 1986; Timms 2006). Although this is mostly due to the greater demand of large hotels (see Chapter 8), foreign ownership and management under neoliberal conditions also contribute to imports (Boxill *et al.* 2004), which weaken linkages with domestic agriculture.

Weakness of marketing

Marketing has been conceptualized as the intermediary actor between supply and demand. The weakness of marketing includes a lack of communication between supply and demand, often attributed to mistrust and misperceptions between producer and consumer (Telfer and Wall 1996; Torres 2003), which arise from social differentiation and resultant power imbalances (Forstner 2004). Furthermore, logistical limitations include poor roads, a dearth of agro-processing to facilitate ease of use and extended storage and lack of cold storage for perishable products (Belisle 1983; Timms 2006).

Although the role of marketing in the linkage process was central to Lundgren's (1973) famous model on linkages, for the most part, it has been overshadowed by a focus on supply and demand constraints. Lundgren's model proposed that over time the dominance of foreign supplies would be eroded by the emergence of local wholesalers who would focus on the marketing of local supplies, in the process stimulating further domestic agricultural production and regional specialization. Many types of local entrepreneurs now exist to serve as marketing intermediaries between supply and demand, including (but not limited to) cooperatives, hawkers/coyotes, supermarkets and large wholesalers. Each have their own set of problems, such as dysfunctional cooperatives (Rhiney 2008; Timms 2006), farmer displeasure with low prices haggled for by hawkers/coyotes (Momsen 1972) and high import content of supermarkets and large wholesalers (Torres 2003).

Again, structural forces in the global economy have played a role in the limitations of marketing domestic products. In the 1970s, the development of the local wholesaler was promoted through import restrictions and government marketing boards, which served the role of linking producers with multiple markets and providing transport and storage facilities (Pattullo 2005; Timms 2006). Neoliberal economic policies in the 1990s disbanded or privatized many government marketing boards as a response to budget cuts and to promote private sector wholesalers. The combination of eased import restrictions and the availability of relatively cheap foreign agricultural products resulted in greater reliance on imports. As an example, Torres (2003) found that at an advanced mass tourism stage, monopolistic wholesalers dominated the hotel supply and sourced products from outside the local region, which runs counter to Lundgren's model.

Contemporary solutions

The global food crisis as opportunity?

Although structural forces emanating from the global scale have contributed to the degradation of domestic agricultural production and increased foreign influence in the tourism industry, with negative repercussions for linkage formation, they can also periodically offer opportunities to proactive actors (Timms 2009). It may sound contradictory, but the current global food crisis provides a strong and, hopefully, sustained incentive for Caribbean nations to finally challenge the legacy of plantation economies by promoting domestic agricultural production as a livelihood opportunity for farmers and food security for the region and to shore up foreign exchange leakages to enhance national economies (World Bank 2008).

Rising food prices can not only stimulate agricultural production but also reorient price-based demand from imports to domestic supplies.

The escalation of food prices is the result of recent restructuring of the global agricultural system. Food prices have, until the present, been undervalued due to subsidized and protected agricultural overproduction in industrialized producers such as the United States and the European Union (Deep Ford and Rawlins 2007; Goodman and Redclift 1990; Timms 2009). Access to markets in developing regions of the world, including the Caribbean, has been facilitated through the previously discussed liberalization of trade policies to alleviate overproduction pressures (Conway and Timms 2003). At the same time, the neoliberal doctrine of comparative advantage refocused Caribbean production on tropical crop exports with food security concerns addressed through the import of cheap (and subsidized) food (Weis 2007).

However, relying on cheap food imports as a food security strategy has proven to be a dangerous one, as global food prices rose dramatically between 2006 and 2008. The contributing factors are many and varied but include a rise in oil prices, which has stimulated the conversion of corn crops towards ethanol production and, in combination with the reconfiguration of shipping routes towards the rising markets of Asia, raised transport costs (Walt 2008). Although past food crises were cyclical and temporary, the World Bank (2008) believes that the contemporary crisis is structural with economic simulations pointing towards the maintenance of high prices in the foreseeable future. If Caribbean farmers can use this sustained incentive to recover domestic food markets, it will serve as a step towards the creation of articulated economies with backward and forward linkages, spreading development in tourism to other sectors of the economy, such as agriculture and beyond.

Strengthening linkages through the four 'Ps' of marketing

With strong incentives to increase domestic agricultural production and disincentives to rely on imports (both related to price and availability), conditions have become more conducive for linkage formation between agriculture and tourism. Taking advantage of this window of opportunity is vital, but to improve the chances of success, we propose adoption of a more inclusive linkage framework that ties supply and demand together through the four 'Ps' of marketing: product, place, promotion and price (Evan-Wong 1996). Re-conceptualizing linkages from this marketing perspective shifts focus from the traditional reactive approach of creating markets for what is already being produced towards a more proactive approach tailoring production to be more responsive to market demands.

This idea relates to the shift that occurred in the field of marketing away from a 'make-and-sell' logic and towards a central focus on the needs of the buyer (Levitt 1960). Recent studies have supported this concept, whereby organizations focused on the interests of the customer outperform those that merely marketed producers' goods (Deshpandé and Farley 2004; Kirca *et al.* 2005). Hence,

Table 7.1 Strengthening agriculture and tourism linkages using marketing's four Ps

	Supply	*Demand*
Product	Quality Reliable selection of locally sourced farm goods	Branding Local, sustainable and authentic branding strategy for region
Place	Logistics Transporting, storing and sorting functions assumed by local intermediaries	Retailing The delivery mechanism for consumption of locally sourced farm goods
Promotion	Relationship marketing Mutually satisfying, sustainable relational exchanges based on communication and trust	Advertising Promotional campaign that communicates local, sustainable and authentic brand
Price	Costs Cost containment of locally sourced farm goods	Value-based pricing Pricing strategy based on how customer perceives value

marketing starts with customer needs and develops ways of creating, delivering and communicating the benefits that satisfy customer needs. This is operationalized as the four 'Ps' of marketing: product, place, promotion and price (McCarthy 1960). At the firm level, the unique combination of this marketing mix is designed to satisfy the target customer. We use this framework to describe the connection between supply and demand and the mechanisms by which agriculture and tourism linkages are ultimately strengthened (see Table 7.1).

Product

The product is the heart of the marketing mix and consists of the bundle of benefits offered and everything the buyer receives in an exchange. Products include goods, services and even ideas, which are translated into features, packaging, brands and many other strategic decisions (Lamb *et al.* 2008). When customers need a job done, they essentially hire the product (Christensen *et al.* 2005). From the customer's point of view, the brand comes to represent the producer's promise as to how the job will be accomplished and can serve to differentiate the product from competitors.

For tourists, the product is the experience, which includes an expectation for local food. Access to quality farm goods locally presents an opportunity to enhance the customer experience. Quality demands reliable availability of a selection of farm goods, which can be further differentiated by designating the source. Branding the product as local, sustainable, organic, fair and/or authentic is responsive to customer needs and may enhance the tourist's experience. This type of product demand based on differentiation and branding plays to possible strengths of local producers but requires flexibility and diversification. In the end, the customer benefits not just through satiety but also by connecting to the place, supporting sustainable farming and contributing to the livelihood of the region.

Conway and Timms (2010) argue for the rebranding of alternative tourism in the Caribbean as 'slow tourism', related to the slow food movement that focuses on local products and stresses quality of experience, and product, over quantity. Such branding is generally seen in more upscale hotels and restaurants, such as Jake's in Treasure Beach, Jamaica, where locally caught fish and produced vegetables are purchased and the daily menu is altered in accordance. As this locale brands itself as community based, the inclusion of local food creates a positive feedback beneficial to all parties. At this point, such practices are mostly limited to smaller enterprises, but there is potential to apply this type of branding to larger hotels and restaurants that may be adopting more flexible specialization (Poon 1990).

Place

Place describes when, where and how benefits are delivered to the customer. It includes the channel through which value exchanges take place between supply and demand (Momsen 1998; Timms 2006). For producer and buyer, the channel performs specialized functions, provides contact efficiency, and helps overcome discrepancies in terms of time, place, quantity and assortment (Coughlan *et al.* 2006). By performing transactional (contacting and negotiating), logistical (transporting and storing) and facilitating (research and financing) functions, channel intermediaries add value to markets (Lamb *et al.* 2008). For suppliers, the channel provides access to markets. For buyers, the channel allows for the acquisition and consumption of need-satisfying products.

On the supply side, the place includes picking up, transporting, delivering and storing the product (such as in cold storage). Place can also include sorting the product into desirable grade and quantity (agro-processing). On the demand side, place is delivery of the product when, where and how the customer demands it and usually takes place on-site at the restaurant where consumption occurs. However, it is important to note that past research on agriculture and tourism linkages in the Caribbean, and elsewhere, have focused almost exclusively on hotel restaurants (Belisle 1984; Telfer and Wall 1996; Torres 2003; Timms 2006). When studying all-inclusive resorts such an approach is satisfactory as the majority of the tourist food consumption will occur at the hotels' restaurants. However, in other cases, significant places that should be included in future research are restaurants that cater to tourists, yet are not associated with a specific hotel (Momsen 1972, 1986). Examples could include large foreign-owned enterprises such as Margaritaville in Montego Bay, Jamaica, or, conversely, smaller locally owned enterprises such as Little Ochie in Alligator Pond, Jamaica. Comparing and contrasting their procurement patterns would be an important addition to the agriculture and tourism linkage literature and practice. Momsen found little difference between hotels and restaurants in St Lucia and Montserrat in their procurement patterns (Momsen 1972, 1986).

Promotion

To satisfy demand, buyers need to be aware of a product's existence, how to acquire and use the product and other informational needs. This is the function of the promotional aspect of the marketing mix. Promotion includes advertising, personal selling and other forms of communication (Lamb *et al.* 2008). To establish and maintain relationships, promotion serves to inform and persuade buyers that a specific product will satisfy an unmet need. A practical example is provided by Torres (2002), where a positive correlation was found between tourist knowledge of local foods and propensity to consume local foods. By contributing to customer knowledge, promotion can reinforce positive brand association and promote usage of locally sourced farm goods (Kotler and Keller 2006).

Although promotion is commonly assumed to be focused purely on the demand side through building awareness and stimulating interest among customers, it also serves a supply-side function by coordinating efforts and maintaining relationships. Communication is key to build commitment and trust among channel members; for example, farmers and hotels or restaurants (Mohr and Nevin 1990; Morgan and Hunt 1994). In this way, promotion supports coordination among agricultural linkages and can help break down the social differentiation between what are viewed as elite chef/tourist and traditional farmer; a major cause of distrust and misperceptions about local products in terms of quality and familiarity (Torres 2003).

Demonstrations, trade shows and other face-to-face promotions can help communicate demand to producers, who can then tailor their production to existing and newly identified and communicated demands by the chef and tourists, who would also be more familiar with local products. For example, the annual Denbigh Agricultural Show, put on by the Jamaica Agricultural Society, includes celebrity cook-offs using local ingredients to promote local agricultural products. Inviting representatives of the tourism industry would be an advantageous way to stimulate more linkages between these two sectors of the economy. On a regional scale, two groups that need to be brought together are the Caribbean Farmers Network and the Caribbean Tourism Organization, based on the mutual advantages such a relationship would foster.

Price

Price represents what the buyer gives up in the exchange. Price can also serve as an important signal for value as buyers may look to price to set expectations about what will be received in the exchange. In addition to monetary costs, buyers assume non-monetary costs associated with acquiring and consuming the product. Marketing must understand what the customer gives up to maximize value. In some cases, customers look to lower costs by seeking inexpensive alternatives, whereas in other cases, customers will trade higher financial costs for added value, thus paying a premium. Marketing's role is to offer the right price for the customer. Although economic and competitive forces are strongly influential, marketing must balance these considerations in determining or adjusting price.

This understanding of pricing consideration is key to foster agricultural and tourism linkages. As with any element of the marketing mix, the producer and seller can tailor their pricing strategy to meet the needs of different segments of customers. Using perceived-value pricing, price is based on what value means to the customer (Zeithaml 1988). For example, discounting is necessary when value means paying a low price for lower perceived quality and quantity. However, customers may be willing to trade higher prices for increased benefits or reduced non-monetary costs, such as organics for the health conscious and fair trade for the socially conscious as is occurring in the United Kingdom (Momsen 2008). Tourists from Europe are more likely to look for such goods while in the Caribbean than are tourists from North America. In addition, higher quality produce can command a premium price targeted at such consciously motivated customers, tourists or hotels, while perceived subpar produce could be sorted and marketed at a lower price to other markets more interested in pure price points (Gibson 1994).

On the supply side, pricing concerns are driven by acquisition, production and delivery costs with a clear need to maintain profit goals to grow the enterprise. With the rise in global food prices, and currency devaluations making imports more expensive, it may become easier for local producers to compete with imported products on price (Torres 2003). However, this is contingent on Caribbean farmers either substituting more expensive imported agricultural inputs with cheaper local inputs or adopting organic production methods (Timms 2009).

Conclusion

In this chapter, we sought to accomplish three main goals. The first was to explain how neoliberal economic policies have contributed to the supply, demand and marketing constraints faced in linking tourism and agriculture in the Caribbean. Second, we argued that the contemporary food crisis resulting from global agricultural restructuring has created a strong incentive for increased domestic agricultural production and more locally oriented tourist demand, which taken together provide an opportunity to strengthen tourism–agriculture linkages. Third, we introduced a framework for understanding linkages from a marketing perspective that we believe is more inclusive of the relationships between supply and demand. Through a brief introduction to the four 'Ps', we hope to stimulate new thought and open up future research in linking tourism and agriculture in the Caribbean.

A number of possible avenues of future research are suggested in this chapter. One is the inclusion of other 'places' where linkages occur beyond hotels, such as other restaurants that cater to tourists. Does age, ownership or size of establishment affect the propensity to purchase locally produced goods in a similar manner as hotel restaurants (Momsen 1998)? Another prospective avenue of research, and practice, would be to evaluate the impact of farm-to-table trade shows, community events and alliances and governmental policy on strengthening agriculture and tourism linkages. As mentioned previously, the Denbigh Agricultural Show in Jamaica would be an excellent site for promoting agriculture and tourism linkages, with research to ascertain its effectiveness. We also believe that more research on the role of intermediaries is called for, including wholesalers and local entrepreneurs, in the linkage process (see Torres 2003; Timms 2006). Finally, the influence of customer perceptions on agriculture and tourism linkages is an important area of inquiry (see Torres 2002).

One possibly controversial recommendation from this chapter is the shift of traditional marketing arrangements from finding markets for what is already produced and instead orienting production to meet existing demands. We believe the former approach is the result of a practical and scholarly focus on agricultural development. While we share this same end goal, we believe it will be better served by transforming domestic agriculture into more diversified and flexible production to serve the demands of the buyers. Through marketing activities, firms are able to translate inputs into customer benefits by offering lower priced and/or higher value goods and services. Although opportunities and threats will continue to evolve due to broader environmental forces, it is marketing systems that will adapt to these changes and ultimately increase the availability and quality of goods and services from which consumers and producers benefit.

However, this illuminates the important caveat that tourism cannot be considered the sole solution for agricultural development; meeting demand from supermarkets and wholesalers, among others, would serve to diversify and expand marketing opportunities. Furthermore, strong agriculture and tourism linkages would promote the formation of a cluster of related industries (Porter 1990), which create spill over effects for nearby industries. For example, demand for goods and services that support agriculture such as fertilizer and transport might be stimulated through increasing linkages between local farmers and tourism service providers. Such a cluster would stimulate conditions whereby firms can rely on key local resources and innovate based on the linkages with demanding customers, sophisticated rivals and supporting institutions such as university and trade associations. Government policy can encourage agriculture and tourism linkages by supporting the formation of such clusters. Ultimately, success may hinge on the development of marketing systems down to the firm level, which requires that enterprises adopt the marketing concept and manage its marketing mix of place, product, promotion and price.

Diversified products and markets will increase and strengthen linkages and lead to more articulated economies with economic benefits spreading from sector to sector. It is this type of development that theorists from Sir Arthur Lewis to the Plantation School have long been calling for. We believe that the marketing mix is one avenue of research and action that is yet to be fully developed and can help achieve these ultimate goals, particularly in Caribbean nations where the nature of small island societies facilitate local interaction (Conway 1998).

References

Beckford, G. (1972) *Persistent Poverty: Underdevelopment in Plantation Economies of the Third World*, New York: Oxford University Press.

Belisle, F.J. (1983) 'Tourism and Food Production in the Caribbean', *Annals of Tourism Research*, 10: 497–513.

Belisle, F.J. (1984) 'Tourism and Food Imports: The Case of Jamaica', *Economic Development and Cultural Change*, 32(4): 819–42.

Best, L. (1976) *The Choice of Technology Appropriate to Caribbean Countries*, Montreal: McGill University Centre for Developing-Area Studies.

Boxill, I., Singh, D.R. and Segree, M.D. (2004) 'Caribbean Tourism and the FTAA', in N. Karagiannis and M. Witter (eds) *The Caribbean Economies in an Era of Free Trade*, Aldershot, UK: Ashgate.

Brohman, J. (1996) 'New Directions in Tourism for Third World Development', *Annals of Tourism Research*, 23(1): 48–70.

Bryden, J.M. (1973) *Tourism and Development: A Case Study of the Commonwealth Caribbean*, Cambridge, England: Cambridge University Press.

Christensen, C.M., Cook, S. and Hall, T. (2005) 'Marketing Malpractice: The Cause and the Cure', *Harvard Business Review*, 83(12): 74–83.

Conway, D. (1998) 'Microstates in a Macroworld', in T. Klak (ed.) *Globalization and Neoliberalism: The Caribbean Context*, Lanhem, MD: Rowman and Littlefield Publishers.

Conway, D. and Timms, B. (2003) 'Where is the Environment in Caribbean Development Theory and Praxis?', *International Development Planning Review*, 3(1-2): 92–130.

Conway, D. and Timms, B. (2010) 'Re-Branding Alternative Tourism in the Caribbean: The Case for 'Slow Tourism'', *Journal of Tourism and Hospitality Research*, 10(4): 329–44.

Coughlan, A.T., Anderson, E., Stern, L.W. and El-Ansary, A.I. (2006) *Marketing Channels*, 7th edn, Upper Saddle River, NJ: Pearson Prentice Hall.

Deep Ford, J.R. and Rawlins, G. (2007) 'Agricultural Trade Policy and Food Security in the Caribbean: Structural Issues, Multilateral Negotiations and Competitiveness', in J.R. Deep Ford et al. (eds) *Trade Policy, Trade and Food Security in the Caribbean*, Rome: FAO.

Deshpandé, R. and Farley, J.U. (2004) 'Organizational Culture, Market Orientation, Innovativeness, and Firm Performance: An International Research Odyssey', *International Journal of Research in Marketing*, 21(1): 3–22.

Dwyer, L. and Forsyth, P. (1993) 'Assessing the Benefits and Costs of Inbound Tourism', *Annals of Tourism Research*, 20(4): 751–68.

Evan-Wong, S. (1996) 'Marketing Agricultural Information Services in the Eastern Caribbean', *Library Management*, 17(3): 22–8.

Fisher, D. (2004) 'The Demonstration Effect Revisited', *Annals of Tourism Research*, 31(2): 428–46.

Forstner, K. (2004) 'Community Ventures and Access to Markets: The Role of Intermediaries in Marketing Rural Tourism Products', *Development Policy Review*, 22(5): 497–514.

Gibson, E. (1994) *Sell What You Sow: A Guide to Successful Produce Marketing*, Auburn, CA: New World Publishing.

Gomes, A.J. (1993) 'Integrating Tourism and Agricultural Development', in D.J. Gayle and J.N. Goodrich (eds) *Tourism Marketing and Management in the Caribbean*, New York: Routledge.

Goodman, D. and Redclift, M. (1990) 'The Farm Crisis and the Food System: Some Reflections on the New Agenda', in T. Marsden and J. Little (eds) *Political, Social and Economic Perspectives on the International Food System*, Aldershot, UK: Avebury.

Hall, M.S. and Shaples, L. (2003) 'The Consumption of Experiences or the Experience of Consumption? An Introduction to the Tourism of Taste', in M. Hall et al. (eds) *Food Tourism Around the World: Development, Management and Markets*, Oxford: Butterworth-Heinemann.

Harker, T. (1989) 'The Caribbean in the Context of the Global Crisis', in G. Beckford and N. Girvan (eds) *Development in Suspense: Selected Papers and Proceedings of the First Conference of Caribbean Economists*, Kingston, Jamaica: Friedrich Ebert Stiftung.

Hope, K.R. (1986) *Economic Development in the Caribbean*, New York: Praeger.

Karagiannis, N. (2004) 'Dependence, Cumulative Causation and the Caribbean', in N. Karagiannis and M. Witter (eds) *The Caribbean Economies in an Era of Free Trade*, Aldershot, UK: Ashgate.

Kirca, A.H., Jayachandran, S. and Bearden, W.O. (2005) 'Market Orientation: A Meta-analytic Review of Its Antecedents and Impact on Performance', *Journal of Marketing*, 69(2): 24–41.

Kotler, P. and Keller, K.L. (2006) *Marketing Management*, 12th edn, Upper Saddle River, NJ: Pearson Education.

Lamb, C.W., Hair, J.F. and McDaniel, C. (2008) *Marketing*, 9th edn, Mason, OH: Cengage Learning.

Levitt, T. (1960) 'Marketing Myopia', *Harvard Business Review*, 38(4): 45–56.

Lewis, W.A. (1955) *The Theory of Economic Growth*, London: Allen and Unwin.

Lundgren, J.O.J. (1973) 'Tourist Impact/Island Entrepreneurship in the Caribbean', in R. Momsen (ed.) *Geographical Analysis for Development in Latin America and the Caribbean*, Chapel Hill, NC: Clag Publications.

McCarthy, E.J. (1960) *Basic Marketing: A Managerial Approach*, Homewood, IL: Irwin.

McElroy, J.L. (1986) 'The Tourism Demonstration Effect in the Caribbean', *Journal of Travel Research*, 25(2): 31–4.

Manley, M. (1982) *Jamaica: Struggle in the Periphery*, London: Writers and Readers Publishing Cooperative Society.

Mohr, J., and Nevin, J.R. (1990) 'Communication Strategies in Marketing Channels: A Theoretical Perspective', *Journal of Marketing*, 54(4): 36–51.

Momsen, J.H. (1972) *Report on Vegetable Production and the Tourist Industry in St. Lucia*, Alberta, Canada: Department of Geography, University of Calgary.

Momsen, J.H. (1986) *Linkages Between Tourism and Agriculture: Problems for the Smaller Caribbean Economies*, Seminar Paper No. 45, Department of Geography, University of Newcastle Upon Tyne.

Momsen, J.H. (1998) 'Caribbean Tourism and Agriculture: New Linkages it the Global Era?', in T. Klak (ed.) *Globalization and Neoliberalism: The Caribbean Context*, Lanhem, MD: Rowman and Littlefield Publishers.

Momsen, J.H. (2008) 'Fair Trade versus Unfair Trade: The Windward Island Banana Industry' *Caribbean Geography*, 15(2):130–41.

Morgan, R.M. and Hunt, S.D. (1994) 'The Commitment-Trust Theory of Relationship Marketing', *Journal of Marketing*, 58(3): 20–38.

Organization of American States (OAS) (1984) *Tourism and Agricultural Linkages in the Caribbean*, Washington, DC: International Trade and Tourism Division.

Pattullo, P. (2005) *Last Resorts: The Cost of Tourism in the Caribbean*, 2nd edn, London: Latin America Bureau.

Poon, A. (1990) 'Flexible Specialization and Small Size: The Case of Caribbean Tourism', *World Development*, 18(1): 109–23.

Porter, M.E. (1990) *The Competitive Advantage of Nations*, New York: Free Press.

Rhiney, K. (2008) 'Towards a new model for improved tourism-agriculture linkages? The case of two farmers' co-operatives in Jamaica', *Caribbean Geography*, 15(2): 142–59.

Stiglitz, J.E. (2003) *Globalization and its Discontents*, New York: W.W. Norton.

Telfer, D. and Wall, G. (1996) 'Linkages Between Tourism and Food Production', *Annals of Tourism Research*, 23(3): 635–53.

Timms, B. (2006) 'Caribbean Agriculture-Tourism Linkages in a Neoliberal World: Problems and Prospects for St. Lucia', *International Development Planning Review*, 28(1): 35–56.

Timms, B. (2009) 'Development Theory and Domestic Agriculture .in the Caribbean: Recurring Crises and Missed Opportunities', *Caribbean Geography*, 15(2): 101–17.

Torres, R. (2002) 'Toward a Better Understanding of Tourism and Agriculture Linkages in the Yucatan: Tourist Food Consumption and Preferences', *Tourism Geographies*, 4(3): 282–306.

Torres, R. (2003) 'Linkages Between Tourism and Agriculture in Mexico', *Annals of Tourism Research*, 30(3): 546–66.

Torres, R. and Skillicorn, P. (2004) 'Montezuma's Revenge: How Sanitation Concerns may Injure Mexico's Tourist Industry', *Cornell Hotel and Restaurant Administration Quarterly*, 45(2): 132–44.

Walt, V. (2008) 'Food Prices: Hunger Strikes', *Time*, June 5, 2008.

Weis, T. (2004) 'Restructuring and Redundancy: The Impacts and Illogic of Neoliberal Agricultural Reforms in Jamaica', *Journal of Agricultural Change*, 4(4): 461–91.

Weis, T. (2007) 'Small Farming and Radical Imaginations in the Caribbean Today', *Race and Class*, 49(2): 112–17.

Williams, E. (1970) *From Columbus to Castro: The History of the Caribbean 1492-1969*, New York: Random House.

Witter, M. (2004) 'Prospects for Jamaica's Economic Development in the Era of the FTAA', in N. Karagiannis and M. Witter (eds) *The Caribbean Economies in an Era of Free Trade*, Aldershot, UK: Ashgate.

WorldBank (2008) 'What are the facts about rising food prices and their effect on the region?' Online. Available at: http://go.worldbank.org/CJYWKZPMX0 (accessed 17 June 2008).

Zeithaml, V.A. (1988) 'Consumer Perceptions of Price, Quality, and Value: A Means-End Model and Synthesis of Evidence', *Journal of Marketing*, 52(3): 2–22.

8 Agritourism linkages in Jamaica: case study of the Negril all-inclusive hotel subsector

Kevon Rhiney

Introduction

Since the 1970s, the tourism industry has grown phenomenally throughout the Caribbean, becoming one of the leading sectors both in terms of job creation and foreign exchange earnings (Mullings 2004; Nurse 2007). Many states in the region have now turned to tourism as their main way of achieving macroeconomic growth and development. In 2008, the region[1] attracted approximately 22.9 million tourists, generating over US$ 27 billion in revenue (Caribbean Tourism Organization 2008). This heavy and growing reliance on the sector has resulted in the Caribbean being increasingly referred to as the most tourism-dependent region in the world (Duval and Wilkinson 2004; Daye *et al.* 2008; Jayawardena 2002; Mather and Todd 1993). According to the World Travel and Tourism Council's (WTTC) Tourism Satellite Accounts data, Caribbean export earnings from international visitors and tourism goods are expected to generate 15.7 per cent of total regional exports (US$ 23.5 billion) and contribute some 12.3 per cent (US$ 39.4 billion) to the region's GDP in 2010 (WTTC 2010).

The problem is that this rapid expansion in regional tourism has generally occurred alongside high rates of foreign exchange leakages and the formation of weak and unplanned linkages with other sectors within host economies (Karagiannis 2004; Phillips and Graham 2007; Ramjeesingh 2006; Timms 2006; see also Chapter 6). In fact, one of the defining features of the Caribbean tourism industry is its heavy reliance on imported goods and services, especially food (Hayle 2005; Momsen 1998).

In the context of Jamaica, tourism growth has long hinged on maximizing visitor numbers as against, and at times, even at the expense of the development of the domestic farming sector. Tourism's phenomenal growth in Jamaica, since its formal inception in the late 1800s (Taylor 1993), has occurred alongside a declining agricultural sector evident through the chronic increase in the sector's negative trade balance over the past three decades. The island's food import bill more than doubled between 1991 and 2001 from US$ 199 million to US$ 503 million (Planning Institute of Jamaica 1992, 2002) and has continued an upward trend ever since. By the end of 2008, Jamaica's food import bill stood at US$ 886

Table 8.1 Distribution of expenditure of stopover tourists, 2008

Categories	*US$ per person per night*	*Percentage of total*
Accommodation (including food and beverages)	68.46	57.3
Food and beverage only	8.60	7.2
Entertainment	12.42	10.4
Transportation	6.93	5.8
Shopping	11.71	9.8
Miscellaneous	11.35	9.5
Total	119.47	100

Source: Jamaica Tourist Board (2008).

million, whereas total food exports amounted to only US$ 264 million – a deficit of US$ 623 million (Planning Institute of Jamaica 2009).

Presently, Jamaica is considered one of the leading tourism destinations in the Caribbean, particularly when measured in terms of visitor arrivals and gross expenditure (Caribbean Tourism Organization 2008; Dodman and Rhiney 2008). In 2008, Jamaica hosted a total of 2.8 million visitors, generating some US$ 1.98 billion in gross visitor expenditure (Planning Institute of Jamaica 2009). That same year, stopover visitor expenditure on food and beverages alone accounted for 7.2 per cent of total tourist expenditure, amounting to US$ 32 million (Table 8.1). Note that this is underestimated because food and beverages are also included within accommodation expenditures, especially in the case of all-inclusive hotels (Dodman and Rhiney 2008).

Food and beverage sales to the tourism market can therefore provide a potentially lucrative source of income for Jamaican farmers, and make a significant contribution to total foreign exchange earnings, once the majority of foods consumed are not imported. Notwithstanding, research findings indicate that while enormous opportunities exist, particularly among large all-inclusive hotels to stimulate local food production in Jamaica, such opportunities have hitherto failed to materialize. Instead, the market linkages between tourism and local agriculture in Jamaica remain weak and disorganized.

The empirical research presented in this chapter builds on the findings of previous studies, particularly the works of Belisle (1980, 1983) and Momsen (1972, 1974) in terms of how hotel characteristics such as ownership, size and class influence hotel food-purchasing patterns. The chapter explores the operations of all-inclusive hotels in Negril and their impact on local agricultural production. This is done against the backdrop of recent developments, including the increasing popularity and dominance of the all-inclusive hotel subsector in the Caribbean (Issa and Jayawardena 2003) and recent claims that the linkages between tourism and agriculture have been strengthening in more mature Caribbean destinations such as Jamaica amidst new demands from tourists for more exotic and rarer varieties of fruits and vegetables (Conway 2004; Momsen 1998; Pattullo 2005).

Table 8.2 Selected indicators for the three major resort towns, Jamaica, 2008

Area	*Accommodation (rooms)**	*Direct employment*	*Stopover arrivals by intended area of stay*
Montego Bay	9,307	11,650	474,755
Ocho Rios	8,202	9,357	441,398
Negril	7,574	9,503	351,404
Percentage share	84.2	86.5	71.7
Total Jamaica	29,794	35,257	1,767,271

* Accommodation includes all types of establishments including guest houses and villas.
Source: Jamaica Tourist Board (2008).

The Negril case study

Most of Jamaica's tourism is confined to the north coast and three resort towns (Montego Bay, Ocho Rios and Negril) in particular. Together, these three resort towns accounted for more than three quarters of the island's available rooms and direct employment and over two thirds of stopover arrivals in 2008 (Table 8.2).

Negril is located on the western tip of Jamaica and is home to a resident population of 5,670 (Statistical Institute of Jamaica 2001).[2] An isolated and remote fishing village until the late 1960s, Negril has expanded to become one of Jamaica's major resort towns – hosting an estimated 351,404 stopover visitors in 2008, representing approximately 20 per cent of Jamaica's total (Jamaica Tourist Board 2008). Negril currently has one of the highest concentrations of hotels and guesthouses per hectare in Jamaica and is the most rapidly expanding of the island's three major resort towns. Between 1998 and 2008, Negril's hotel room capacity increased by approximately 34 per cent – more than five times that of Montego Bay, the island's leading resort town (Figure 8.1).

Negril provides a unique opportunity for the forging of viable market linkages between tourism and local agriculture in Jamaica. Unlike other major resort towns such as Montego Bay and Ocho Rios, there is no established local produce market in the area. The bulk of the food supplied to Negril is therefore a response to tourists' food demand and is supplied directly to hotels and restaurants as there is no intervening opportunity in the linkages formed.

Hotels in Negril obtain their food supplies from several major sources. These sources include large wholesale companies, small contractors or intermediaries, small retail outlets and supermarkets, individual contract farmers, farmers' associations and several nearby produce markets – the most frequented being the Savanna-la-mar produce market located approximately 30 km from the town in the parish of Westmoreland. Hotels may rely on several of these sources to meet their food needs, and these may be both local and imported.

Negril is also renowned for its eclectic mix of large-scale contemporary mass tourism development (epitomized by the recent opening of a number of all-inclusive properties) existing alongside smaller scale traditional type hotels, guesthouses and restaurants. Despite their limited numbers, all-inclusive hotels

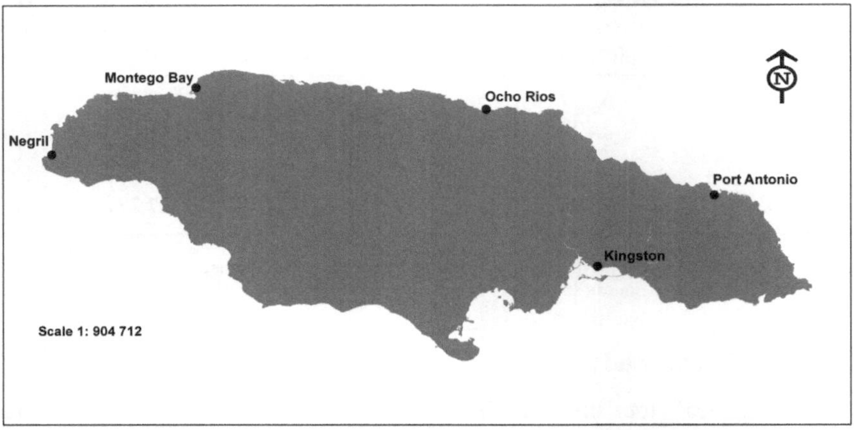

Figure 8.1 Major resort towns in Jamaica.

dominate Negril's tourism economy. Although amounting to less than one quarter of hotels in the study area, all-inclusives accounted for 54 per cent ($N = 2{,}798$ rooms) of the total bedroom stock, with a mean of 215 bedrooms per hotel. This assortment of hotels (particularly in terms of the differences in size and operation modes) allows for a comparative analysis of the ways these various establishment types operate and their resulting ability to stimulate local agricultural production.

Methodology

The empirical data presented in this chapter are part of a broader study focusing on tourism and agriculture linkages in Jamaica. The chapter draws on an extensive survey of 44 of Negril's hotels, representing 62 per cent of all the hotels in operation during the time the study was carried out. The survey extended over approximately 6 weeks between July and August 2005 and involved a wide cross section of industry stakeholders, including hotel chefs, property owners, food and beverage managers and food purchasing managers.

The questionnaire survey was designed specifically to identify trends in local food purchasing patterns among different categories of hotels, guesthouses and restaurants operating in Negril. Some of the major areas of focus for the survey included the identification of the major suppliers of local and imported foodstuff to the area, problems encountered by hotel representatives when sourcing local foods, menu information, including dish specialization, and the nationality and training experience of executive chefs. The survey also looked at business proprietors' attitudes towards the purchasing and consumption of locally grown food provisions as opposed to food imports, including their level of confidence in the ability of local farmers to meet their food demand quota and quality requirements.

For purpose of analysis, hotels were categorized into three broad groups based on size, ownership (local, foreign or combined) and room price. Hotel size was measured based on the number of bedrooms available per establishment. Initially, this was divided into three categories: small (less than 50 rooms), medium (50–99 rooms) and large hotels (100 rooms or more). However, given the extremely skewed nature of the data collected, hotels were recoded in two rather than three categories with regards to size – small (less than 50 rooms) and large (50 rooms and over) – to allow for a more effective analysis of the data.

Negril is characterized by a large number of small hotels translating into a small percentage of the total hotel room stock and a small number of large hotels accounting for more than one half of the total number of hotel rooms (Dodman and Rhiney 2008; Larsen 2008). Of a total of 44 establishments that offered accommodation for tourists, 29 (65.9 per cent) were classified as small hotels accounting for 769 rooms or approximately 23 per cent of the total room stock of all the hotels surveyed in the census and 15 (34.1 per cent) were classified as large hotels, amounting to 2,609 bedrooms or approximately 77 per cent of the total room stock. The range in hotel size is large; the smallest hotel has only three rooms and the largest hotel has 420 rooms. Because of the skewed data distribution, the mean hotel size is 77 rooms, whereas the median size is 31 rooms.

Room costing was measured by the average daily cost for a single bedroom per unit. Less than US$ 50 was referred to as low-cost accommodation, US$ 50–99 as medium-cost and US$ 100 or more per night as high-cost. Room costing served as a proxy to Belisle's 'class' variable (see Belisle 1980: 32), as it was believed to be more suitable in describing the differences in accommodation pricing, given Jamaica's present situation. The fact that a larger share of the high-cost accommodations surveyed were found offering all-inclusive packages implies that these hotels can neither be categorized as 'high-class' nor even 'high-cost' because their principal *modus operandi* usually revolves around offering inexpensive vacation packages (Enoch 1996). In the Jamaican context, the industry is characterized by the promotion of high-volume package tours from North America (cf. Boxill 2004), unlike the more high-end clientele typical of other Caribbean destinations such as Anguilla, St Maarten and the Turks and Caicos (Randall 2006).

A food inventory was later carried out to ascertain the quantity of food (local and imported) being consumed in the area by both value and volume. This inventory was contingent on the establishments having a good record of their food purchasing data. As it was discovered that a large number of the smaller establishments were unable to contribute to this exercise due to insufficient record keeping, a stratified random sample was taken of 22 hotels (15 small and 7 large properties) that had initially participated in the survey and were found to have good food purchasing records. This method of sampling when applied to strata of significantly different size, such that the smaller are oversampled relative to the larger, allows for statistical comparisons to be made between the two groups (Parfitt 2005). In the case of the Negril sample frame, there were nearly twice as many small hotels than large hotels.

Tourism and local food production in Jamaica

Like many other Caribbean islands, early tourism development in Jamaica was largely premised on its ability to benefit other sectors of the local economy, especially agriculture, through the forging of backward and forward linkages (Zinder & Associates 1969). Tourism's expansion was expected to stimulate domestic agricultural production in Jamaica by encouraging local farmers to both expand and diversify their cropping systems to meet the industry's growing demand (Lundgren 1973). However, a number of scholars (Bryden 1974; Belisle 1980) writing in the 1970s have in fact shown that the tourism industry then provided only limited opportunities for domestic agriculture. In 1968, Jamaican hotels were reportedly importing some 69.4 per cent of their food supplies (Cazes 1972: 145).

Other studies carried out throughout the period reflected more or less the same trend – that agriculture–tourism linkages were weak. Brown (1974: 137) estimated that Jamaican hotels were importing between 66 and 79 per cent of their food and beverage needs. Although the bulk of these food imports consisted of meats and meat products (some 42 per cent of the total food imports), fruits and vegetables accounted for a significant proportion of the total. A few years later, Belisle (1980: 75) reported that the percentage consumption of food imports by Jamaican hotels had declined to 54 per cent. Belisle (1980) linked this decrease in food imports to the import-substitution policies employed by the Jamaican government during the 1970s.

However, Belisle (1980) discovered that although considerable import-substitution had been attained, formal linkages between the two sectors remained weakly developed. Although Belisle highlighted a variety of factors inhibiting the two sectors from forging linkages between each other such as inconsistencies in food supply, quality and pricing, hotel characteristics such as size and class were pointed out as being major causes for the absence of linkage between the two sectors. Belisle observed, for instance, that large and high-end/high-class hotels displayed a greater tendency to purchase food imports than small and low-end/low-class hotels.

Belisle (1980: 137) further explained that high-class hotels purchased more imported foodstuffs due to the willingness of their clientele to pay higher prices for better quality and greater variety of food, whereas larger hotels found it easier to use prepared, precooked and portion-controlled foods – which were almost exclusively imported – given their larger customer base. An underlying assumption in Belisle's study was that linkages between tourism and local agriculture would improve through capacity-building initiatives such as the transfer and adaptation of appropriate production technologies, land redistribution, expansion of extension services and agricultural research and the continued enforcement of import-substitution policies.

Since Belisle's research, only a few other studies (Organization of America States 1994, 1997; Rickard and Carmichael 1995) have been carried out in Jamaica that looked at the nature and extent of the linkages that exist between tourism and local agriculture. These studies have, however, only partially looked at the relationship between the two industries. Despite this paucity of information, a number of scholars have surmised that market linkages between the two sectors have since improved in larger island destinations such as Jamaica attributed mainly to changing attitudes towards serving local cuisine in hotels, the globalization of food consumption habits, a maturing of regional tourism industries and increased government assistance to domestic producers (Conway 2004; Momsen 1998; Pattullo 2005).

The problem as hinted at before is that these claims are yet to be substantiated by empirical evidence. Although Rickard and Carmichael (1995) revealed that Jamaican hotels had increased their use of local food suppliers and that a more complex supplier network had indeed developed, with local foods being sourced from farming communities located further inland, more recent evidence suggests that the majority of foodstuffs consumed by Jamaican hotels is still being imported (Dodman and Rhiney 2008; Pennicook 2006). The most recent estimate suggests that the Jamaican tourism industry experiences a 50 per cent foreign exchange leakage rate, the majority of which is attributed to the import of food (Ramjeesingh 2006). These arguments are very much attuned with the findings of this research, as fieldwork data indicate that existing market linkages between domestic agriculture and the Jamaican tourism industry remain weak and filled with numerous challenges.

The problem is twofold, extending to both ends of the supply chain. For the agricultural industry, the main problem lies with the domestic farming sector's inability to guarantee sufficient supplies of high-quality, competitively priced agricultural produce on a consistent basis. Limitations range from poorly organized food marketing and food production systems to unfavourable agro-ecological and climatic conditions, infrastructural deficiencies, competition from lower priced food imports, limited use of technology and poor access to well-needed technical and financial support (Barker and Beckford 2008; McGregor *et al.* 2009; Rhiney 2008, 2009). These limitations pose a severe constraint not only for menu planning but also for the creation and sustenance of viable market linkages between the two sectors.

These marketing and production deficiencies have been made worse by the ongoing market liberalization of the Jamaican economy since the 1990s that has effectively made it more profitable to trade in food imports than in local produce (Beckford *et al.* 2007; Weis 2004). This competition coupled with prohibitive production costs has prompted many local farmers to cease producing certain types of crops that are deemed uncompetitive. Presently, the majority of onions sold in Jamaican produce markets and to the tourism industry are sourced overseas from countries such as China and Holland. In contrast, onion production in Jamaica has decreased from 2,457 tonnes in 1998 to a mere 215 tonnes in 2007 – a decline of over 90 per cent.

For the tourism sector, the problem seems to be a lack of communication between industry representatives and local farmers, particular hotels' high food supply quota and quality requirements and a preference by certain types of hotels to source cheaper priced food imports over local food provisions. Large all-inclusive hotels in particular displayed a greater tendency to consume food imports compared with other types of hotels. This supports findings from two studies published by the OAS in 1994 and 1997, which showed all-inclusive hotels in Jamaica to be larger consumers of imported products than non-all-inclusive properties. Both studies concluded that although all-inclusive hotels generated the largest amount of revenue, they had less impact on the Jamaican economy per dollar of revenue than other accommodation subsectors. Missing from these surveys, however, was an adequate explanation of the particular reasons shaping all-inclusive properties' propensity to consume greater volumes of imported foodstuffs relative to locally grown provisions.

Given the subsector's increasing importance in the Caribbean (Issa and Jayawardena 2003; Larsen 2008), a greater understanding of the particular ways the operations of all-inclusive hotels may or may not stimulate local agricultural production can help to elucidate the underlying causes for the characteristically high food import content of many regional tourism industries.

The Negril all-inclusive hotel subsector and its link with local agriculture

Jamaica's tourism industry is largely represented by the remarkable growth of a few large all-inclusive resorts such as the Super Clubs and Sandals chains of hotels (Boxill 2004; Issa and Jayawardena 2003) and more recently, the spate of Spanish-owned all-inclusive hotels such as the RIU, Iberostar and Fiesta groups. When measured against performance-based indicators such as visitor expenditure, occupancy rates and length of stay, all-inclusive hotels are by far the leading contributors to the Jamaican tourism industry (Table 8.3), generating higher GDP, employment and foreign exchange per room than other types of hotels (Issa and Jayawardena 2003). In 2008, there were 60 all-inclusive resorts in Jamaica representing over two thirds of the island's total hotel room stock (Jamaica Tourist Board 2008: 81).

All-inclusive hotels are, however, renowned worldwide for their isolationist mode of operation as evident in destinations such as the Dominican Republic (see Coles 2004; Freitag 1994; Pattullo 2005) and Cancún, Mexico (see Torres 2002), where they predominate. In Jamaica, all-inclusive hotels are credited for the continued growth of the Jamaican tourism industry in recent years, despite the island's reputation of having one of the highest crime rates in the world per capita (Alleyne and Boxill 2003; Issa and Jayawardena 2003). All-inclusive hotels are noted for attracting visitors through their promise of a safe and secure accommodation, one that shields and protects guests from the perceived crime problems that exist in the wider Jamaican society (Boxill 2004). The fenced-off compounds, guarded gates and secluded locations of most of the all-inclusive

Table 8.3 Summary of Jamaica's main tourism indicators, 2000–2008

Indicators	*2000*	*2001*	*2002*	*2003*	*2004*	*2005*	*2006*	*2007*	*2008*
Total visitor arrivals (000)	2,230.3	2,116.9	2,131.8	2,482.9	2,514.6	2,614.5	3,015.4	2,880.3	2,856.2
Total stopovers (000)	1,322.7	1,276.5	1,266.4	1,350.3	1,414.8	1,478.7	1,678.9	1,700.8	1,767.3
Hotel room nights sold (000)	2,956.3	2,903.6	2,830.6	2,971.9	3,134.0	3,253.1	3,572.2	3,653.3	3,722.3
Average hotel occupancy (%)	58.6	56.3	55.5	57.9	61.4	61.9	62.8	63.2	60.4
All-inclusive occupancy (%)	70	69.4	66.2	69.3	71.9	70.0	71.2	71.2	67.9
Non-all-inclusive occupancy (%)	39	35.1	38.3	38.5	39.1	42.4	42.3	46.6	43.1
Total visitor expenditure (US$ million)	1,332.0	1,233.0	1,209.0	1,351.0	1,436.6	1,545.1	1,870.6	1,910.0	1,975.5

Source: Jamaica Tourist Board (various years).

properties seen in Negril seemingly support this argument. Anecdotal evidence from taxi and store operators in the study area suggests that, in some instances, guests were being discouraged by hotel staff from taking local taxis or from even venturing outside the hotels' premises.

Another more subtle reason for visitors' lack of willingness to venture outside these properties relates to the all-inclusive package itself, with its emphasis on a 'cashless experience' for the tourist. As the visitors have already paid for their hotel services, there is little or no incentive for them to venture outside, as this would mean losing out on paid-for goods and services (Hawkes and Kwortnik 2006). When guests do venture outside the property, they usually do so only on escorted excursions to selected sites and gated shopping districts.

As the bulk of the money spent by consumers on all-inclusive packages is restricted to the confines of the hotel property, the consumption of more local food products by these hotels represents one of the most viable options for destination countries such as Jamaica to reduce the high degree of foreign exchange leakages associated with this form of tourism development. However, to date, efforts to maximize the economic returns from Jamaica's tourism industry continue to be focused on increasing visitor numbers, whereas linkages with the local farming sector remain weak and unplanned. If market linkages between the two sectors remain weak, the expected increase in visitor arrivals to Jamaica (Ministry of Tourism and Sports 2001) will simply result in the import of more agro-food items. In light of this, the chapter now turns to the particular reasons that can best explain the greater tendency for all-inclusives to consume imported as opposed to locally grown foodstuffs in the context of Negril.

The food inventory that was carried out during the survey revealed that all-inclusive properties accounted for some 80 per cent of the food by volume consumed by hotels in Negril and 78 per cent by value. The seven all-inclusive hotels included in the inventory were recorded as consuming as much as 30,049 kg of food per week, amounting to J$ 2.92 million by value (approximately US$ 41,000) compared with only 7,103 kg of food per week for the 15 non-all-inclusive hotels. In addition to this, these large all-inclusive hotels purchased a much wider variety of food items when compared with non-all-inclusive resort properties (Table 8.4). Fruits and vegetables accounted for a significant share of the hotels' operation costs. Of a sample of 30 selected fruits and vegetables, the 22 hotels sampled consumed around 37,152 kg per week, amounting to J$ 3.76 million.

Judging solely by the food inventory data, it would seem that large all-inclusive hotels have a greater potential for stimulating local food production than other hotel types, given that the majority of foods being consumed are sourced locally. This is particularly important for Jamaica, in light of the current efforts to find new means of enhancing tourism's contribution to the national economy (Boxill 2004; Dodman and Rhiney 2008; Ramjeesingh 2006;) and in the context of a declining agricultural sector (Beckford and Barker 2007; Rhiney 2008, 2009; Timms 2008). Despite being unable to quantify the average proportion of local to imported foodstuffs consumed by hotels in the study area, largely due to

Table 8.4 Food inventory of Negril hotels by value and volume

Food item	*Quantity purchased (kg)*	*All-inclusives (kg)*	*Non-all-inclusives (kg)*	*Total cost (JMD)*
Tomato	3,397	2,650	747	436,726
Cabbage	2,001	1,298	703	148,250
Cucumber	1,399	978	421	100,554
Cauliflower	786	680	106	94,969
Spinach	208	200	8	9,280
Carrot	2,037	1,490	547	188,999
Squash (yellow)	1,041	1,034	7	136,700
Onion	2,115	1,590	525	141,030
Broccoli	388	358	30	52,510
Sweet pepper	1,377	1,121	256	228,192
Iceberg lettuce	1,145	1,100	45	158,563
Romaine lettuce	988	930	58	157,992
Irish potato	4,261	3,170	1,091	249,955
Sweet potato	664	620	44	46,650
Yam	1,263	1,110	153	118,214
Ginger	84	50	34	13,313
Garlic	271	210	61	34,866
Green beans	114	70	44	13,255
Okra	109	105	4	11,625
Pumpkin	820	610	210	49,763
Beet	97	95	2	7,045
Papaw	2,890	2,240	650	175,612
Watermelon	5,639	4,650	989	461,433
Pineapple	3,611	3,250	361	296,535
Zucchini	447	440	7	52,400
Plantain	NA	NA	NA	50,832
Ripe bananas	NA	NA	NA	160,284
Lime	NA	NA	NA	102,283
Orange	NA	NA	NA	57,120
Total	37,152	30,049	7,103	3,754,950.0
Percentage	100	80.9	19.1	

Note: A number of the items could not be quantified based on the way they are usually procured. For example, plantains, bananas, limes and oranges were mostly purchased by the dozen. As such the author was unable to quantify these into kilograms.

establishments' failure and lack of interest of hotels in recording such information, analysis of the fieldwork data illuminated several distinct patterns that could aid in explaining all-inclusive hotels' greater propensity to consume food imports than other non-all-inclusive properties.

The first reason lies with how price sensitive all-inclusive hotels were found to be when compared with other hotels. Large all-inclusives were found to be particularly sensitive to fluctuations in food supply and food prices. As such, food sourcing operations tended to be price driven and extensive, involving a wide range of suppliers. Torres (2003) showed in the case of Mexico that this need to minimize production costs generally translates into greater local purchases from all-inclusives for key products needed in high volume. However, this is the reverse in the case of small developing states such as Jamaica and indeed the wider Caribbean, where imported food can be obtained at lower prices than local agricultural provisions. Indeed, only approximately 40 per cent of the industry representatives that were surveyed rejected the statement that local foods are generally more expensive than food imports, the overwhelming majority of whom were from small non-all-inclusive properties.

Local farmers' inability to guarantee competitive prices for domestic food crops is partly embedded in traditions that have characterized Caribbean agriculture since slavery, particularly its agricultural monoculture landscape, which has long hindered the introduction of alternative crops as well as limited plans for expanding and diversifying the domestic farming sector (Beckford 1968; Mintz 1985; Momsen 1998). Jamaica's domestic farming sector is dominated by small farmers operating on marginal inland landholdings (Barker 1993). The majority of these farmers operate in isolation of each other, with little or no direct state support and are constrained by inadequate technical assistance, limited access to funding and appropriate technology.

The problem of food price is further compounded by the enormous emphasis that all-inclusives' place on lowering their production costs. Most of Negril's all-inclusive hotels now employ a tendering system where suppliers bid for short-term contracts. The bidding system applies a downward pressure on food prices because food merchants are forced to compete with each other based on price. In most instances, intermediaries opt to source cheaper food imports in an effort to secure the market and maintain their profit margins. Others force local farmers in need of a market to sell their produce for unprofitable sums.

When hotel representatives were questioned about their level of satisfaction with food purchasing in general, two distinct trends were identified. Approximately 78 per cent of the non-all-inclusive hotels indicated that they were generally satisfied with the local food network, compared with only 25 per cent of the all-inclusive subgroup. Representatives from the all-inclusive hotels generally held the view that local producers are incapable of adequately meeting their food quota and quality requirements on a consistent basis. Concerns ranged from local farmers' limited use of technology, to poor post-harvest handling, disorganized marketing networks and inconsistent and unfavourable weather conditions.

The discontent all-inclusive representatives expressed towards the efficiency of the local food network to meet their food quota requirements is best explained by the nature of their food catering operations. Aside from their price consciousness, all-inclusive hotels' higher demand for food, both in terms of quantity and consistency of supply, make them extremely vulnerable to localized food shortages. The majority of these hotels have to rely on a large and complex network of small independent intermediaries to satisfy their supply needs. In this particular arrangement, a tourist establishment communicates its needs in advance to a purveyor at a set price. Food is then sourced from a large number of farming

Figure 8.2 Farming communities identified supplying fresh produce to Negril.

communities spanning the entire island (Figure 8.2). In most cases, no formal contract exists between the food establishment and the supplier. Instead, relationships are primarily based on verbal agreements that can be terminated at any point. Each supplier competes with his contemporaries in a free market environment.

Despite accounting for the bulk of local foods supplied to Negril, the particular marketing arrangement proved to be highly unreliable and filled with myriad problems and challenges extending throughout the entire local food chain. It was found that not only were intermediaries supplying a significant amount of food imports to the area but their operations were also deemed as posing a major barrier to the forging of direct links between local farmers and food proprietors (Rhiney 2008, 2009). As agricultural produce is usually sold fresh with little or no value added, local farmers benefit the least from this arrangement as they are structurally placed at the bottom of the value chain. Furthermore, this system of marketing, where food is supplied by an intermediary, creates a structural disconnect between the producer and the end-user. This means that neither party is knowledgeable about nor is capable of effectively responding to the needs of the other. In short, this preponderance of intermediaries is therefore not a function of tourism's improved linkages with domestic agriculture as suggested by Rickard and Carmichael (1995) but rather the failure of both sectors to forge more viable and sustainable linkages with each other.

Problems associated with the seasonality of certain local food crops and the susceptibility of the local farming sector to a number of agro-climatic hazards also contributes to the general distrust, particularly among expatriate workers, of the capacity of the local food network to meet their food quota and quality requirements on a regular basis. Natural hazards such as hurricanes and droughts are particularly damaging to the local farming sector. Hurricanes are among the most frequent and potentially devastating hazards affecting the Jamaican agricultural sector in recent years (see McGregor *et al.* 2009). The hurricane season officially starts in July and ends in November of each year – just before the start of Jamaica's tourist winter season. Hurricanes therefore pose a significant threat to the island's agricultural production and to the forging of sustained linkages between the tourism and agriculture industries. Between the 2004 and 2005 hurricane seasons, Jamaica was affected by four hurricanes (Ivan, Charley, Dennis and Emily) and tropical storm Wilma, amounting to J$ 9.6 billion in damage to the island's agricultural sector (Planning Institute of Jamaica 2005, 2006). After a hurricane, most suppliers have to resort to imports to maintain their markets. During this time, farmers find it particularly difficult to re-enter the supply chain due to the lengthy period it usually takes for them to recover and the cost of recovery itself.

Another plausible explanation for all-inclusives recording a greater dissatisfaction with the local food network could be their propensity to employ more foreign-born workers than other types of hotels. All-inclusive properties were found employing a smaller percentage of Jamaicans in food management positions than non-all-inclusives. Although more than three quarters (81 per cent) of the non-all-inclusive hotels had more than 80 per cent Jamaican food management staff, only one third (33 per cent) of the all-inclusive hotels employed this level of local staff.[3] The survey also revealed that more all-inclusive hotels employed chefs with either foreign nationalities or overseas training than other hotel types (Table 8.5). The majority of these chefs came from either North America or Europe, but others came from countries as varied as Mexico, Dominican Republic, Trinidad, Japan and Israel. This finding lends support to previous studies that have shown that foreign chefs are more likely to express dissatisfaction with local products and are less inclined to prepare local dishes compared with local chefs (see, e.g. Momsen 1998; Pattullo 2005; Torres 2003)

Fieldwork data confirmed that the hiring of foreign-born and overseas-trained chefs may have serious implications for the incorporation of indigenous speciality food items in hotel menus. Interviews revealed that foreign-born and foreign-trained chefs were less willing to incorporate local speciality items such as yam, sweet cassava and okra in their menus. These chefs generally complained about product quality, primarily produce appearance and size. In addition, they often complained about sanitation standards and seasonality of local food products. A number of the chefs pointed out that some local vegetables such as lettuce and green cabbage are often infested with pests and that they usually end up discarding a significant portion of these. Instead, the majority of foreign chefs preferred to use frozen or processed food imports or food varieties they are familiar with and know how to prepare.

When asked about the type of dishes served, a distinct relationship was also observed. Approximately two thirds of the hotels surveyed stated that they specialized in both local and international cuisine – this was the case for both types

Table 8.5 Differences in the food preparation characteristics among hotels (Values for percentage hotels recorded in each category)

Mode of operation	*Chef nationality*			*Chef training*		*Dish speciality*		
	Non-Jamaican	*Jamaican*	*Dual*	*Local*	*Overseas/ combined*	*Local*	*International*	*Both*
All-inclusive properties	51	41	8	25	75	—	31	69
Non-all-inclusive properties	2	98	—	81	19	18	14	68

of hotels. However, although less than one fifth of the non-all-inclusive hotels specialized only in the provision of local dishes, none of the all-inclusive hotels were found specializing solely in Jamaican dishes. The data therefore suggest that all-inclusive hotels are more likely to specialize in international cuisine and non-all-inclusive hotels in Jamaican cuisine. All-inclusives generally provide a variety of cuisines to their customers, including Italian, Japanese, Tex-Mex (which is a cross between Texas and Mexican cuisines), Continental, Asian and Mediterranean cuisines alongside Jamaican/Caribbean speciality dishes, the latter are usually served on specified theme nights each week.

It is interesting to consider the factors that explain why large all-inclusive resorts market themselves in these particular ways and how these in turn shape the type of foods sourced by these hotels. Although cognizant that international cuisine does not necessarily equate to the use of imported food products and *vice versa*, all-inclusives' preference for international dishes over local dishes is indicative of the standardized ways they are marketed globally and the externally oriented manner in which they operate.

The fact that all-inclusive hotels tend to employ fewer locals in food management, in particular chefs with local training, and also specialize more in international as opposed to local cuisine, has far reaching implications for Jamaica's local food economy. Large all-inclusive hotel chains such as Sandals and Riu have a greater global reach than smaller hotels and tend to standardize their product offerings to reduce production costs. This mirrors results found by Torres (2003), who suggests that the overwhelming emphasis placed on Fordist-type packages by these large-scale mass tourism developments means that less focus is placed on promoting indigenous speciality dishes. In the case of Negril, various local foods such as yams, dasheens and okra are not promoted to tourists. Instead, a substantial portion of these foods are served to staff members rather than guests in staff canteens usually situated in the 'back stages' of these hotels.

The tremendous emphasis placed on promoting 'internationally diverse' as opposed to local indigenous foods by all-inclusive hotels is depicted by the following quote drawn from the website of one of Negril's renowned all-inclusive properties:

> After hours of playing in the tropics, ravenous appetites are satisfied by savory meals – any time of the day or night. For food lovers of all ages, five world-class restaurants fulfill every culinary fantasy with a world of choices.

On the same website, the hotel offers a picture of one of its five international restaurants. The picture shows a family as they are entertained by a seemingly local chef who is dressed in Japanese attire while preparing a meal. Although the images and texts depicted on the website speak of the ways Jamaica's tourism offerings are shaped by larger scale global processes, it also sheds light on the mode or type of tourism model being pursued by the Jamaican government – one that is based on the promotion of high-volume package tours from North America (Boxill 2004; Mains 2008).

Mass tourists have been noted to be among the most conservative with regard to food preferences (see Torres 2003). Therefore, although tourism service providers aim at offering a distinctively unique product, the highly competitive and unpredictable nature of today's international tourism market means that these efforts have to be framed to meet international quality and safety standards (Mains 2008). Often this can have significant material consequences particularly as it relates to the culinary experiences of tourists (Dodman and Rhiney 2008). The continued domination of the Jamaican economy by this conventional model of mass tourism implies that substantially expanding tourists' consumption of locally grown food items might prove to be difficult.

However, despite the growing dissatisfaction with this (large-scale conventional) form of tourism (Boxill 2004; Hayle 2005; Dodman and Rhiney 2008; Ramjeesingh 2006), recent efforts by the Jamaican government to promote smaller scale alternative forms of tourism has been lacking (Mains 2008). Jamaica, like many other developing countries faced with a high debt burden and characterized by a fragile and weakening economy, has tended to pursue an economic strategy that favours the setting up of these large-scale mass tourism developments (Britton 1991; Pantin 1999; Roberts 2008). Often this economic strategy is pursued to the detriment of some local stakeholders, either directly, as is the case with small tourism enterprises (Dahles 1999), or indirectly, as seen in the case of craft market vendors in Jamaica (Boxill 2004).

Despite the inability to forge an extensive link with the local farming sector, a few all-inclusive hotels have recently made attempts to stimulate local agricultural production. A case in point is the Sandals Farmers Programme, initiated in part by the Sandals all-inclusive chain in Jamaica. The programme began in 1996 and entailed a two-pronged approach between the all-inclusive chain and several small farmer co-operatives located throughout the island. The arrangement was that the particular hotel chain would supply seeds to the farmers on a crop-lien basis – whereby seeds were provided to the farmers on credit who in turn made the necessary loan repayment to the hotel at the end of the cropping season. The main objectives of the programme were to improve the quality of local foods supplied to the hotels, diversify and expand the local cropping systems to suit the tourism industry's food demand, develop a proper pricing system and improve the communication between the two groups (Meyer *et al.* 2004; Rural Agricultural Development Authority 2003; Sharpley and Telfer 2008).

Besides the fact that this arrangement does not exist for the majority of all-inclusive hotels operating in Jamaica, I have shown elsewhere that the Sandals Farmers Programme has reaped mixed results (Rhiney 2009). In fact, nearly 14 years have elapsed since its establishment, and only three of the six farmer groups initially involved in the programme are still functional.

Towards a more sustainable agritourism strategy

This study reveals that although opportunities exist for large all-inclusive hotels to stimulate local food production in Jamaica, these have generally failed to materialize. In fact, all-inclusive hotels were found to be significant consumers of food imports. Although the tendency for all-inclusive properties to rely heavily on lower priced imports to satisfy their food needs was linked to them having greater property sizes and a wider customer base, the nature of the all-inclusive concept itself was also found to play a major role in the subsector's general predisposition to consume greater quantities of food imports compared with other hotels. All-inclusives' greater tendency to specialize in international cuisine, employ fewer Jamaicans in food management positions and employ chefs with foreign nationalities and overseas training lend support to this assertion. Added to this are the many production and marketing deficiencies associated with the local farming sector.

Although the factors limiting the intersectoral linkage between tourism and agriculture are many and multifaceted, they are not incurable and are partly symptomatic of the type of policies (or the lack thereof) undertaken by successive Jamaican governments and the ways in which both sectors continue to operate in isolation of each other. The emphasis for the tourism ministry continues to be on improving the industry's economic returns by increasing visitor numbers and mobilizing foreign direct investment for large-scale development projects, while sectoral linkages with local agriculture remain weak. Without a commensurate improvement in the market linkages between the two sectors, an increase in visitor numbers will simply result in the import of more agro-food products to the island.

A number of studies have emphasized a need to develop alternative forms of tourism that will both expand visitor stay and expenditure, thus increasing the industry's net retention capacity (Dodman and Rhiney 2008; Hayle 2005). Implicit in these studies is the view that conventional models of mass tourism are characterized by uneven forms of development and provide little benefit to host economies. Although this may be true, it is doubtful if conventional mass tourism will ever be fully replaced by alternative models – particularly in the case of small island states such as Jamaica, which have become increasingly dependent on mass tourism. As a result, it is imperative that these host economies adopt strategic ways of increasing the net returns from this already existing conventional form of tourism.

One such strategy includes increasing the use of locally grown food products in mass tourists' food consumption practices. Operators of all-inclusive properties should be actively encouraged to adopt more flexible menus that take into consideration the seasonal availability of various local fruits and vegetables. A series of forums and workshops could also be set up to facilitate dialogue between stakeholders in both industries on product requirements and standards and other related marketing issues that are deemed necessary if market linkages are to be formed and maintained. The establishment of standards and the creation of formalized contracts can help in bridging the communication gap between primary stakeholders in both industries and ameliorate the instability in food prices for local agricultural products. The sorting, grading, labelling and packaging of locally grown agricultural products should make up a significant part of this initiative.

If stronger and more sustained ties are to be developed between the two sectors, emphasis also has to be placed on boosting the productive capacity of local farmers, thereby increasing domestic food production as well as improving the way local foods are supplied and marketed. As the bulk of Jamaica's farmers operate on a small scale, the setting up of farmer groups can play a major role in empowering and equipping these farmers to respond effectively to changes associated with the new global environment and the demand requirements of the tourism industry (Rhiney 2009). Farmers' associations not only allow for small farmers to pool their resources but also provide an effective means of transferring technologies to the local agricultural sector and allow for the establishment of direct and greater communication between farmers and tourism representatives. This not only helps to reduce the mismatch between food supply and demand but also aids in alleviating farmers' dependence on intermediaries.

The integration of local food production with tourists' food consumption in Jamaica can indeed provide an important stimulus for the expansion and diversification of domestic agriculture and the retaining of much-needed foreign exchange. For this to become a reality though, there needs to be an overhauling of the two sectors to better complement each other as well as the forging of strategic partnerships between the various stakeholders and interest groups involved. The provision of local goods and services should make up an integral component of the local tourism industry and not just accommodation and entertainment. Likewise, tourism should be positioned as a viable niche market for locally grown food produce and their by-products.

Notes

1 These figures include data from Cancún (Mexico), Cozumel (Mexico) and the Hispanic Caribbean.

2 The last population census for the area was published in 2001. The next census is due for publication in 2011.

3 Chi-square results confirmed that a relationship does exist between the two variables ($x^2 = 0.005$ [df = 6], $p < 0.05$), though a fairly weak correlation (contingency coefficient = 0.425).

References

Alleyne, D. and Boxill, I. (2003) 'The impact of crime on tourists arrivals in Jamaica', *International Journal of Tourism Research*, 5(1): 381–91.

Barker, D. (1993) 'Dualism and disaster on a small tropical island: some constraints on agricultural development in Jamaica', *Tidjschrift voor Econimische en Sociale Geographi*, 84(1): 332–40.

Barker, D. and Beckford, C. (2008) 'Agricultural intensification in Jamaican small-scale farming: vulnerability, sustainability and global change', *Caribbean Geography*, 15(2): 160–70.

Beckford, G. (1968) 'Toward an appropriate theoretical framework for agricultural development, planning and policy', *Social and Economic Studies*, 17(2): 233–42.

Beckford, C. and Barker, D. (2007) 'The role and value of local knowledge in Jamaican agriculture: adaptation and change in small-scale farming', *The Geographical Journal*, 173(2): 118–28.

Beckford, C., Barker, D. and Bailey, S. (2007) 'Adaptation, innovation and domestic food production in Jamaica: some examples of survival strategies of small-scale farmers', *Singapore Journal of Tropical Geography*, 28(1): 273–86.

Belisle, F.J. (1980) *Hotel food supply and local food production in Jamaica: A study in Tourism Geography*, Unpublished doctoral dissertation. London: University Microfilms International.

Belisle, F.J. (1983) 'Tourism and food production in the Caribbean', *Annals of Tourism Research*, 10(4): 497–513.

Boxill, I. (2004) 'Towards an alternative tourism in Jamaica', *International Journal of Contemporary Hospitality Management*, 16(4): 269–72.

Britton, S. (1991) 'Tourism, capital and place: towards a critical geography of tourism', *Environment and Planning Development: Society and Space*, 9(4): 451–78.

Brown, H. (1974) 'Impact of the tourist industries on the agricultural sectors: the competition for resources and the market for food provided by tourism, the case of Jamaica', *Proceedings of the Ninth West Indian Agricultural Economics Conference*, 9(1): 129–42.

Bryden, J. (1974) 'The impact of tourist industries on the agricultural sectors', *Proceedings of the Ninth West Indian Agricultural Economics Conference*, 9(1): 153–62.

CaribbeanTourism Organization (CTO) (2008) *Caribbean Tourism Statistical Report, St. Michael*, Barbados: Caribbean Tourism Organization.

Cazes, G. (1972) 'Le Rôle du Tourisme Dans la Croissance Economique: Reflexions a Partir de Trois Examples Antillais' ['The role of tourism in economic growth: Reflections from three examples in the Antilles'], *The Tourist Review*, 27: 93–8, 144–8.

Coles, T. (2004) 'What makes a resort complex? Reflections on the production of tourism space in a Caribbean resort complex', in D.T. Duval (ed.) *Tourism in the Caribbean: Trends, Development, Prospects*, London: Routledge, pp. 235–56.

Conway, D. (2004) 'Tourism, environmental conservation and management and local agriculture in the eastern Caribbean: is there an appropriate, sustainable future for them?', in D.T. Duval (ed.) *Tourism in the Caribbean: Trends, Development, Prospects*, London: Routledge.

Dahles, H. (1999) 'Tourism and small entrepreneurs in developing countries: A theoretical perspective', in H. Dahles and K. Bras (eds) *Tourism and Small Entrepreneurs: Development, National Policy, and Entrepreneurial Culture: Indonesian Cases*, New York: Cognizant Communications.

Daye, M., Chambers, D. and Roberts, S. (2008) 'Caribbean tourism: New perspectives', in M. Daye, D. Chambers and S. Roberts (eds) *New perspectives in Caribbean tourism*, London: Routledge, pp 1–16.

Dodman, D. (2004) 'Postmodernity or Profitability? Modes of Tourism in Jamaica', in S. Courtman (ed.) *Beyond the Blood the Beach and the Banana: New Perspectives in Caribbean Studies*, Kingston: Ian Randle Publishers.

Dodman, D. and Rhiney, K. (2008) '"We Nyammin?" Food Supply, Authenticity and the Tourist Experience in Negril, Jamaica', in M. Daye, D. Chambers and S. Roberts (eds) *New Perspectives in Caribbean Tourism*, London and New York: Routledge.

Duval, D. and Wilkinson, P. (2004) 'Tourism development in the Caribbean: Meaning and influence', in D. Duval (ed.) *Tourism in the Caribbean: Trends, Development, Prospects*, London: Routledge.

Enoch, Y. (1996) 'Contents of tour packages: A cross-cultural comparison', *Annals of Tourism Research*, 23(3): 599–616.

Freitag, T.G. (1994) 'Enclave tourism development: for whom the benefits roll?' *Annals of Tourism Research*, 21(3): 538–54.

Hayle, C. (2005) 'Tourism in Jamaica: The Impact of the Past on the Future', in C. Jayawardena (ed.) *Caribbean Tourism: Visions, Missions and Challenges*, Kingston and Miami: Ian Randle Publishers.

Hawkes, E. and Kwortnik, R.J. (2006) 'Connecting with the culture: a case study in sustainable tourism', *Cornell Hotel and Restaurant Administration Quarterly*, 47(4): 369–81.

Issa, J. and Jayawardena, C. (2003) 'The "all-inclusive" concept in the Caribbean', *International Journal of Contemporary Hospitality Management*, 15(3): 167–71.

Jamaica Tourist Board (JTB) *(various years) Jamaica: Annual Travel Statistics*, Kingston: Jamaica Tourist Board.

Jayawardena, C. (2002), 'Mastering Caribbean tourism', *International Journal of Contemporary Hospitality Management*, 14(2): 88–93.

Karagiannis, N. (2004) 'Dependence, Cumulative Causation and the Caribbean', in N. Karagiannis and M. Witter (eds) *The Caribbean Economies in an Era of Free Trade*, Ashgate: Aldercroft.

Larsen, S.C. (2008) 'Negril in the News: content analysis of a contested paradise', *Caribbean Geography*, 15(1): 35–58.

Lundgren, J. (1973) 'Tourist impact/island entrepreneurship in the Caribbean', in J. Momsen (ed.) *Geographical Analysis for Development in Latin America and the Caribbean*, Chapel Hill: CLAG Publications.

Mains, S. (2008) 'Island village, tourist world: Representing cultural geographies of Jamaica', *Caribbean Geography*, 15(1): 1–13.

Mather, S. and Todd, G. (1993) *Tourism in the Caribbean*, Special Report No. 455, London: The Economist Intelligent Unit.

McGregor, D., Barker, D. and Campbell, D. (2009) 'Environmental change and Caribbean food security: Recent hazard impacts and domestic food production in Jamaica' in D. McGregor, D. Dodman and D. Barker (eds) *Global Change and Caribbean Vulnerability: Environment, Economy and Society at Risk*, Kingston, Jamaica: University of the West Indies Press.

Meyer, D., Ashley, C. and Poultney, C. (2004) 'Tourism-agricultural linkages: boosting inputs from local farmers', *Business Implementation of Pro-Poor Tourism Series*, Paper 3. Online. Available at: www.pptpilot.org.za/Briefing%20papers/paper3.pdf

Ministry of Tourism and Sports (2001) *The Ten Year Master Plan for Sustainable Tourism Development*, Kingston, Jamaica: Author.

Mintz, S. (1985) *From plantations to peasantries in the Caribbean: Caribbean contours*, Baltimore, MD: John Hopkins University Press.

Momsen, J.H. (1972) *Report on Vegetable Production and the Tourism Industry in St. Lucia*, Calgary: University of Calgary, Department of Geography.

Momsen, J.H. (1974) *Report on Vegetable Production and the Tourism Industry in Montserrat*, Calgary: University of Calgary, Department of Geography.

Momsen, J. (1998) 'Caribbean tourism and agriculture: new linkages in the global era?', in T. Klak (ed.) *Globalization and Neoliberalism: The Caribbean Context*, USA: Rowman and Littlefield Publishers.

Mullings, B. (2004) 'Globalization and the territoralization of the new Caribbean service economy', *Journal of Economic Geography*, 4(3): 275–98.

Nurse, K. (2007) *Science, Technology and Innovation in the Caribbean', presentation, Technology Policy and Development in Latin America*, Santiago, Chile: UNECLAC. Online. Available at: http://vi.unctad.org/files/papers/nursepaper.pdf

Organization of American States (OAS) (1994) *Economic Analysis of Tourism in Jamaica*, Washington, DC: Organization of American States.

Organization of American States (OAS) (1997) *Tourism in Jamaica: An Economic Analysis*, Washington, DC: Organization of American States.

Pantin, D. (1999) 'The challenge of sustainable development in small developing states: A case study on tourism in the Caribbean', *Natural Resources Forum*, 23(4): 221–33.

Parfitt, J. (2005) 'Questionnaire design and sampling', in R. Flowerdew and D. Martin (eds) *Methods in Human Geography: A guide for students doing a research project*, 2nd edn, London: Pearson-Prentice Hall.

Pattullo, P. (2005) *Last Resorts: The cost of tourism in the Caribbean*, Kingston: Ian Randle Publishers.

Pennicook, P. (2006) 'The All-Inclusive Concept: Improving Benefits to the Jamaican Economy', in K. Hall and R. Holding (eds) *Tourism: The Driver of Change in the Jamaican Economy?* Kingston and Miami: Ian Randle Publishers.

Phillips, W. and Graham, C.B. (2007) 'Issues of Agro-Tourism in the Caribbean', in C. Jayawardena (ed.) *Caribbean Tourism: More than Sun, Sand and Sea*, Kingston and Miami: Ian Randle Publishers.

Planning Institute of Jamaica (PIOJ) (various years) *Economic and Social Survey of Jamaica*, Kingston: Planning Institute of Jamaica.

Ramjeesingh, D. (2006) 'Import content of tourism: Explaining differences among Island States', *Tourism Analysis*, 11(1): 33–44.

Randall, R. (2006) 'Eastern Caribbean tourism: developments and outlook', in R. Sahay, D. Robinson and P. Cashin (eds) *The Caribbean: From Vulnerability to Sustained Growth*, Washington DC: International Monetary Fund.

Rhiney, K. (2008) 'Towards a new model for improved tourism-agriculture linkages? The case of two farmers' co-operatives in Jamaica', *Caribbean Geography*, 15(2): 142–59.

Rhiney, K. (2009) '(Re)Defining the link? Globalisation, tourism and the Jamaican food network', in D. McGregor, D. Dodman and D. Barker (eds) *Global Change and Caribbean Vulnerability: Environment, Economy and Society at Risk?*, Kingston, Jamaica: University of the West Indies Press.

Rickard, T. and Carmichael, B. (1995) 'Linkages between the agricultural and tourism system in sustaining rural development in Jamaica', in C. Bryant and C. Marois (eds)

Proceedings of the First Meeting of the IGU Study Group on the Sustainability of Rural Systems, Montreal: Department of Geography, University of Montreal.

Roberts, S. (2008) 'Reflections from the periphery: An analysis of small tourism businesses within the sustainability discourse', in M. Daye, D. Chambers and S. Roberts (eds) *New Perspectives in Caribbean Tourism*, London: Routledge.

Rural Agricultural Development Authority (RADA) (2003) Sandals Farmer Programme. Online. Available at: www.radajamaica.com.jm/marketing_articles.asp?

Sharpley, R. and Telfer, D.J. (2008) *Tourism and Development in the Developing World*, London and New York: Routledge.

Statistical Institute of Jamaica (STATIN) (2001) *Jamaica Population Census 2001*, Kingston: Statistical Institute of Jamaica.

Taylor, F. (1993) *To Hell with Paradise: A History of the Jamaican Tourism Industry*, Pittsburgh and London: University of Pittsburgh Press.

Timms, B. (2006) 'Caribbean agriculture-tourism linkages in a neoliberal world: Problems and prospects for St. Lucia', *International Development Planning Review (IDPR)*, 28(1): 35–56.

Timms, B. (2008) 'Development theory and domestic agriculture in the Caribbean: recurring crises and missed opportunities', *Caribbean Geography*, 15(2): 101–17.

Torres, R. (2002), 'Cancún's tourism development from a Fordist spectrum of analysis', *Tourist Studies*, 2(1): 87–116.

Torres, R. (2003) 'Linkages between Tourism and Agriculture in Mexico', *Annals of Tourism Research*, 30(3): 546–66.

Weis, T. (2004) 'Restructuring and redundancy: the impacts and illogic of neoliberal agricultural reforms in Jamaica', *Journal of Agrarian Change*, 4: 461–91.

World Travel and Tourism Council (WTTC) (2010) 'Caribbean – Key Facts at a Glance'. Online. Available at: www.wttc.org/eng/Tourism_Research/Economic_Research/Regional_Reports/Caribbean

Zinder & Associates (1969) *The future of tourism in the eastern Caribbean*, Washington, DC: United States Agency for International Development (USAID).

9 Tourism and agriculture in Barbados: changing relationships

Pamela Richardson-Ngwenya and Janet Momsen

Introduction

New global food systems and World Trade Organization policies make it increasingly difficult for Caribbean farmers, who face notoriously high-input costs, to compete for both domestic and export markets. Since the loss of preferential trading arrangements for traditional cash crops such as bananas (Clegg 2002) and sugar (Richardson-Ngwenya 2009, 2010), this marginalization has become even more critical. Increasingly, farmers are turning to the production of vegetables, fruits and other food items for the local tourist market. At the same time, higher world food prices are demanding more local food production in net food importers such as Barbados to protect food security and reduce inflation (ECLAC 2008).[1] Tourists bring the market to the farmers and as a result, tourism can be seen as a non-traditional export (Brohman 1996). Across the Caribbean, especially in the smaller islands of the region, tourism is replacing the traditional plantation export agriculture as the main driving force of the economy. In 2008, the region attracted approximately 22.9 million tourists, generating over US$ 27 billion in revenue (Caribbean Tourism Organization [CTO] 2008).[2] The heavy and growing reliance on this sector has resulted in the Caribbean being increasingly referred to as the most tourism-dependent region in the world as noted by the CTO: '[B]y almost any economic indicator, the Caribbean is four times more dependent on the tourism industry than any other region of the world (CTO 1998: 5)'. Caribbean export earnings from international visitors and tourism goods are expected to generate 15.7 per cent of total regional exports (US$ 23.5 billion) and contribute some 12.3 per cent (US$ 39.4 billion) to the region's GDP in 2010 (World Travel and Tourism Council [WTTC] 2010; see Chapter 8). Furthermore, the Caribbean has the largest number of people employed in tourism (25 per cent) and the highest proportion of GDP from tourism when compared with the rest of the world (CTO 2007).

In Barbados, tourism is now seen as the main engine of economic growth (Emmanuel 2010). It is viewed as a long-established destination with visitor arrivals peaking in 2004 at over 2.4 million and declining slightly since then. Arrivals from Europe, mainly from the United Kingdom, rose from 26 per cent in 1980 to 48 per cent in 2000, the highest proportion of British tourists in any

Caribbean destination (Momsen 2004: 280). According to the Ministry of Tourism, the current recession brought only a 10 per cent decline in visitor arrivals to Barbados in 2009, which is less than that in other regional destinations. In 2009, earnings from tourism in Barbados remained much the same as in 2008, with major discounting of unnecessary hotel charges (Thangaraj 2009). The Ministry urged stakeholders in tourism to find creative solutions to the distribution of tourism products and services in a sustainable way (ibid.; Momsen 2006). Fostering more extensive and intensive linkages between the local agro-food sector and tourism has become a key strategy towards achieving this sustainability (Inter-American Institute for Cooperation on Agriculture [IICA] 2008).

This chapter looks at changes in the relationships between domestic agriculture and tourism over the last half-century in Barbados. By rephrasing old questions and bringing together estranged concepts, we hope to provide a new understanding of tourism and agriculture linkages in Barbados. An important perspective that we bring to bear on these linkages draws from the recent agro-food network literature (Goodman 2003; Harris 2010, Marsden and Smith 2005). This literature is broad ranging and interdisciplinary, using and interrogating concepts of globalization, localization, embeddedness, materiality, hybridity, symbolic economies and more, to engage critically with today's changing geographies of food and agriculture (Wilk 1999). A key feature of this literature is an attempt to engage such geographies in a non-binary and integrative way, conceptualizing interactive networks of production, trade, regulation and consumption (Richardson and Whatmore 2009). In contrast, studies of tourism–agriculture linkages are largely data heavy and policy focused, lacking critical theorization. In response, this chapter draws on the agro-food literature, mobilizing the concepts of embeddedness, localization, globalization and symbolic economies to interpret the changing relationships between tourism and agriculture in Barbados. We first outline the historical dimensions of tourism and agriculture on the island, before commenting on the changing relationship between these sectors in light of what can be considered a 'new food economy'. We argue that using an integrative and multiscalar approach is imperative to a wider geographical understanding of tourism–agriculture linkages.

Historical dimensions of tourism–agriculture linkages in Barbados

The globalization and democratization of tourism began in the 1960s with the spread of long-distance air travel. A newly affluent middle class endowed with ever increasing amounts of disposable income and leisure time propelled tourism to its current status as the world's largest industry. With the loss of protected markets for agricultural exports, tourism has become of ever greater importance in small Caribbean islands, as measured by increasing tourist density ratios (Momsen 1998). Caribbean tourism is based on unequal power relations that are manifested at several geographical scales. These can be seen at the world economy level between core nation hotel and travel companies and peripheral host nation tourist boards; at the regional level between small, resource-poor, tourism-dependent islands and larger, wealthier Caribbean nations with more diversified economies; and at the local level between largely poor, black, rural hosts and rich, white, urban guests. Such inequalities have encouraged a focus on studies of Caribbean tourism and its (often negative) impact on local cultures and socio-economic benefits (Bryden 1974; Patullo 1996). In this chapter, with its focus on Barbados, we take a different approach and consider linkages between tourism and other sectors of the economy, which are increasingly being seen as vital to sustainable tourism development (IICA 2008; Momsen 2006). Sustainability in tourism is of increasing interest in the face of climate change (McKercher *et al.* 2010), which in Barbados is affecting the viability of coral reefs and the availability and potability of the water supply (Emmanuel 2010).

Feeding tourists

Over the last half-century, attitudes concerning the linkages between tourism and agriculture have changed. 'Linkage' has been understood primarily in terms of the potential and capacity of local farmers to supply fresh produce to hotels and resort restaurants. Government planners and foreign consultants have long argued that tourism would stimulate other sectors of the Caribbean economy, notably agriculture (Alleyne 1974; Bryden 1974; Gomes 1993; Sergent 1967). Gooding (1971) calculated that although almost two thirds of the food consumed by tourists in 1968 was imported into Barbados, if import duties and taxes, local transport and mark-up to local distributors and wholesalers are included, an impressive 52 per cent of the money spent on tourist food stayed in the island. In studies carried out in the 1970s and 1980s, one of the co-authors found that hotel consumption of local foods had increased in St Lucia and Montserrat as tourist interest in local food grew and government assistance to producers became evident (Momsen 1986). In the 1990s, Gomes (1993) calculated that food and beverages served in Caribbean hotels, including the costs of delivering the meals and drinks to visitors, accounted for 57 per cent of the total revenue per room.

In Barbados, in 1987, only 11 per cent of small farmers sold vegetables and fruits to hotels and they were mainly those based in the west coast tourist areas, but almost three quarters of hotel food was locally produced when items such as soft drinks, eggs, milk and chicken were included (Momsen and Aspinall 1987). In a survey in 1990, it was found that hotels (N = 24) estimated that 46 per cent of operating costs were spent on food and drink, whereas for restaurants (N = 28), the figure was 49 per cent. Both said that over 90 per cent of chicken, eggs, vegetables, milk and soft drinks were local, while demand was highest for additional local vegetables and meat. Among the hotels, 21 (88 per cent) served local foods unfamiliar to foreign tourists, and most hotels and restaurants felt that training chefs to use local foodstuffs was important (Momsen 1990). Hinds (2003) found similar patterns of hotel and restaurant local food consumption with all leaf lettuce, tomatoes, cucumber and most other vegetables sourced locally, with the exception of English potatoes. Fruit is mostly imported from nearby islands, such as St Lucia, whereas most herbs are locally supplied.

Changes in local agriculture and food production

With respect to local food consumption in the 1960s and 1970s, both Alleyne (1974) and Bryden (1974) felt that the situation in Barbados was not so positive. Bryden (1974: 153) predicted that the seasonal competition between tourism and (sugar cane) agriculture for labour and land would lead to a perverse relationship of immiserating conflict in the smaller islands of the region (Torres and Momsen 2004). In the case of Barbados, Alleyne, using data from the late 1960s, felt that there was insufficient land to grow enough vegetables and fruits for tourists, as this activity would compete with land for sugar cane. He also concluded that farmers with small holdings of less than 10 hectares were not interested in growing vegetables for sale.

Repeat surveys conducted over a period of four decades in Barbados (Momsen 2005), have found that farmers' behaviour and land use patterns have changed: in 1963, only 8 per cent of small farms surveyed had more than half their land in vegetables compared with one third of farms in 1987 and 45 per cent of farms in 2003. In 1987, almost two thirds of the farmers interviewed did not have fruit trees or bananas, whereas by 2003, almost all farms had at least some stools of bananas and 5 per cent of farmers said that fruit was their most important crop. So, over 40 years, small farmers in Barbados have gone from being peasant sugar cane producers to specialist vegetable, herb, root crop and fruit growers. New items such as honeydew melon, celery and ornamental plants were also being grown by 2003. With the loss of subsidized markets for sugar in Europe and the consequent closure of all but one sugar factory in Barbados, sugar cane is no longer a profitable crop, especially for small-scale farmers. Indeed, according to the small farmer liaison officer at the Barbados Agricultural Management Company, there were probably only 300 small farmers left in production in 2007, contributing 2.6 per cent of cane production, down from about 15 per cent in 1995 (personal communication, 3 May 2007). In 2008, Barbados produced 31,611 tonnes of raw sugar from 15,268 acres of harvested cane land (Agronomic Research and Variety Testing Unit 2006: 3). This is demonstrative of progressive decline in sugar production, which peaked during the early 1970s. The contribution of sugar to GDP has declined from 21 per cent in 1960 to 6.3 per cent in 1980 to 1.4 per cent in 2005, and acreage under cane as well as the amount of sugar produced by Barbados has fallen by more than 50 per cent since the 1980s. This has opened up space for diversification within agriculture, but it must also be noted that competition for land from non-agricultural sectors, in particular real estate development, has soared. Many sugar plantation owners are selling at least some of their land off for non-agricultural purposes (Richardson-Ngwenya 2009: 192–193).

Although farmers have become more diversified in their crop production, they have also become increasingly part-time, with multiple income streams. In 1987, 35 per cent of farmers got more than half of their income from the sale of farm produce, whereas in 2003, only 27 per cent did. However, in 2003, the average proportion of income coming from such sales was 38 per cent, so although fewer got the majority of their income from sales of farm produce, the average for all farmers was still over one third of income. There was less sharing of produce with friends in 2003 than in 1987, which may indicate both greater commercialization of production and/or more home consumption of local foods.

These changes in farmers' behaviour and land use would seem to be a response to both the loss of the traditional plantation sugar export market and increased demand for land from other sectors. Further contributing factors are the growth in tourist demand for local foodstuffs and the demonstration effect on the local population of tourist eating habits. It is also possible that returning migrants from the United Kingdom and the United States have influenced local consumption habits, fuelling demand for non-traditional crops.

The new food economy in Barbados

The Caribbean rural economy is in the process of being transformed from a food regime based on settler colonialism largely dependent on imported foodstuffs to one in which transnational and regional food networks are reconfigured by the globalization of tourism. A growing proportion of tourists in the Caribbean, especially those attracted by ecotourism and heritage tourism, represent the 'new consumerism' that characterizes the foodscapes of Western countries (Dolphijn 2004; Goodman and DuPuis 2002; Goodman 2004). Such consumerism demands healthy natural food produced locally by a sustainable agriculture that does not damage the environment (Emmanuel 2010). Consumers also want to avoid contaminated food, which the Barbados government sees as a danger inherent in importing so much food (IICA 2008). However, as climate change leads to increasing water shortages in Barbados, the water itself is becoming contaminated by agricultural chemicals and other pollutants thus possibly becoming unsuitable for the irrigation of items containing high levels of water, such as lettuce (Emmanuel 2010). Thus, the need for more organic production and efforts to use more efficient types of irrigation are urgent.

Tourism itself has engendered a commodification of place (Britton 1991) that can be selectively appropriated to develop a saleable commodity. Vacation-based acquaintance with local dishes and foods can generate a market for food and beverage souvenirs. Production of such souvenirs has been a growth industry for many Caribbean agro-industries. Souvenir foodstuffs are usually sold in small amounts in expensive and eye-catching packages so that the unit price is much higher than that in the local market. Typical souvenir food items are specialist rums, brown sugar, coffee, jams, jellies, spices, hot sauce and tropical fruit juices and pastes.

Challenges

There have been several key challenges to the development of linkages between tourism and agriculture in Barbados, from the food supply perspective. First is the tourist demand for non-local produce. Today, this does not represent a lack of interest in local foods but is rather an ironic effect of the new consumerism.

For example, consumer demand for fair trade and organic products cannot usually be sourced locally in Barbados. Furthermore, the way in which symbolic economies configure and commodify place, in particular through the promotion of national/local cuisines and décor (Cook and Crang 1996), means that there is a high demand for Italian, Mexican, French and American cuisines! This presents obvious difficulties to sourcing foods locally for the provision of such restaurants.

Local cultural politics also constitutes an entrenched challenge to the development of supply relationships between local farmers and tourist venues. This is a difficult factor to account for, but interviews have suggested a deep lack of trust between farmers and hotel purchasers (Richardson 2004). The asymmetries of power between farmers and buyers influence both parties' attitudes. The high level of risk associated with small-scale farming is often not appreciated by buyers, who prefer to order from 'safer' sources: large wholesalers and plantations. For small farmers, there is a high level of risk in terms of the environmental factors inherent in farming and sometimes, access to transport for deliveries. Furthermore, the hotel industry is often irregular with its food requirements, which are dependent on guest numbers and needs. Hotels and restaurants also need small amounts delivered often, which is not always appreciated by small farmers who may have to travel a long distance to deliver for a small return. Payment is often made through an invoice system that brings much delay and frustration for small farmers, thus fostering further antagonism. In an atmosphere of mistrust, small farmers are in need of secure contracts from hotels, but hotels want the flexibility to source the best produce available at the best price.

A final challenge to note is the lack of economies of scale and competitive producers in Barbados, in the context of the increasing globalization of food supply chains. In many cases, Barbados cannot compete with imported products in terms of price. As long as the quality of produce is high, price is the bottom line for buyers and is often the most important factor, which undermines the ability of local farmers to sell their produce to hotels. There is a long history of food importation in the Caribbean run by a powerful local merchant class and wholesale sector whose influence cannot be discounted. However, with global food prices soaring in the current economic crisis, this may be the opportune moment for local organic producers, with few (petrochemical-based) input costs, to secure a footing in the tourist market on the basis of both quality and price.

New developments

Today, the demand for agricultural produce from the hospitality industry in Barbados is estimated at about US$ 60 million a year (IICA 2008). Although there is a desire for locally grown farm produce and fresh fish, some of the major limitations of the hotel and restaurant sector's ability to procure such foodstuffs are the inconsistencies in the local supply chain such as availability, quality, price, reliability, logistics and convenience. The recently developed IICA/Barbados Hotel and Tourism Association initiative, the Farm-to-Table Programme, aims to develop an enhanced tourism product with a culinary aspect, with events such as the 'Taste of Barbados' festival, involving top-quality Zagat-rated restaurants and the production of local cookbooks (IICA 2008). Growing awareness of the local diabetes epidemic and its dietary requirements is linking culinary and health tourism. The Government of Barbados is promoting the use of Barbadian foods in hotels and the development of a 'Bajan haute cuisine'. It is also increasing efforts to provide credit, improved seeds and lower prices for irrigation water to farmers to help them increase production of vegetables. Breeding of the native Black Belly sheep to produce a heavier animal yielding more meat is also taking place. However, lack of storage and processing facilities is still a problem. The objective of these official efforts is to guarantee farmers a market for their produce and to expand the export markets for Barbadian produce. There have also been a few recent developments in agritourism, whereby tourists visit farms for recreational, heritage or educational activities (including eating and shopping!), and some minor development in 'farm-stay' style holidays, therefore expanding the local scope of tourism–agriculture linkages.

Conclusion

By taking a multiscalar and integrative approach, considering the significance of globalization and localization and of symbolic economies and the new consumerism, this chapter has offered an original insight into tourism–agriculture linkages and the situation of Barbados in particular. The goal of the new National Strategy is to increase economic development in Barbados by the promotion of 'agri-tourism linkages through the strengthening of inter-sectoral linkages between agriculture and tourism … that have multiple benefits to both sectors and by extension to overall national development' (IICA 2008: 19). This is based on similar linkages already established in Nevis, St Lucia and Jamaica (see Chapter 8 and Rhiney 2008). However, these linkages may be ephemeral when they involve a single hotel, as in Nevis, where in 1990, the Four Seasons Hotel set up an arrangement with a cooperative of farmers to supply fresh vegetables, with the Ministry of Agriculture acting as a middleman (ECLAC 2005). Sales to the hotel grew quickly at first but stabilized after 2003, and then in 2009, the hotel closed down because of the recession (Thangaraj 2009). The much larger hotel base of Barbados should ensure that farm to hotel links are not so dependent on the fortunes of a single hotel.

Growth in tourist consumption of local foods is partly attributable to the globalization of diet and greater familiarity with international cuisine. The hospitality industry is increasingly international, with chefs in many Bajan hotels being trained overseas, though they are West Indian by origin and so familiar with local foods. Today, the food being served to tourists in Barbados is proudly advertised as coming from local sources. Many of the small farmers of Barbados are behaving more like the organic producers of North America than the traditional peasant farmers of 40 years ago. However, at the same time, many of the specialist producers of crops such as tomatoes and lettuce are foreign or returnee migrants, and the entrepreneurs producing food souvenirs are also often expatriates (Richardson 2004).

Community-based tourism as in Dominica and St Lucia where tourists visit farms and take part in farm activities such as the picking and processing of coffee and cocoa and try local foods can be considered agricultural tourism. This spreads the benefits of tourism to a wider range of stakeholders and can increase the demand for local products. In Barbados, the weekend Oistins Fish Fry is the second most popular event among tourists after visits to Harrisons Cave (ECLAC 2005). The Ministry of Tourism is planning to expand tourism in Barbados into farm stays and other programmes of home accommodation to provide a greater variety of offerings to tourists and involve more Barbadians (Thangaraj 2009). Although it appears that tourism *per se* will not enhance the development of the agriculture sector in Caribbean countries, it can facilitate its sustainability through encouragement of organic and fair trade production and tourist demand for and consumption of these products.

Acknowledgements

Janet Momsen would like to thank the students of the University of Newcastle upon Tyne and the University of California, Davis for their assistance in collecting the data in 1987 and 2003 respectively. She would also like to thank Valerie Beynon, Kelly Payson and Travis Marcotte for their help with data collection and analysis.

Notes

1 Food inflation from 2006 to 2008 was 21 per cent compared with 8.5 per cent for non-food inflation (ECLAC 2008).
2 These figures include data from Cancún (Mexico), Cozumel (Mexico) and the Hispanic Caribbean.

References

Alleyne, F. (1974) '*The Impact of Tourism and Its Concomitant Unrealised Potential for Agricultural Development in the Barbadian Economy*', Proceedings of the Ninth West Indian Agricultural Economics Conference, St Augustine: University of the West Indies, 143–52.

Agronomic Research and Variety Testing Unit (ARVTU) (2006) *Sugar Cane Variety Update 2006*, Barbados Agricultural Management Company, Barbados.

Britton, S. (1991) 'Tourism, capital and place: Towards a critical geography of tourism', *Environment and Planning D. Society and Space*, 9: 451–78.

Brohman, J. (1996) 'New directions in tourism for Third World development', *Annals of Tourism Research*, 23: 48–70.

Bryden, J.M. (1974) '*The Impact of the Tourist Industries on the Agricultural Sectors: The Competition for Resources and Food Demand Aspects (With Special Reference to the Smaller Economies of the Region)*', Proceedings of the Ninth West Indian Agricultural Economics Conference, St Augustine: University of the West Indies, pp. 153–62.

Caribbean Tourism Organization (CTO) (1998) *Proceedings of the First Caribbean Hotel and Tourism Investment Conference*, Bridgetown, Barbados: CTO.

Caribbean Tourism Organization (CTO) (2007) *Sustainable Tourism Development Strategy and Action Plan for the Caribbean*, Barbados: CTO.

Caribbean Tourism Organization (CTO) (2008) *Caribbean Tourism Statistical Report, St. Michael*, Barbados: Caribbean Tourism Organization.

Clegg, P. (2002) *The Caribbean Banana Trade: From Colonialism to Globalization*, Palgrave, Macmillan.

Cook, I. and Crang, P. (1996) 'The world on a plate - Culinary culture, displacement and geographical knowledges', *Journal Of Material Culture*, 1(2): 131–53.

Dolphijn, R. (2004) *Foodscapes: Towards a Deleuzian Ethics of Consumption*, Delft: Eburon.

ECLAC (2005) 'Caribbean Tourism and Agriculture: Linking to enhance Development and Competitiveness', LC/CAR/L.76. December 21st.

ECLAC (2008) 'The escalation in world food prices and its implications for the Caribbean', LC/CAR/L.179. Online. Available at: www.eclac.cl/publicaciones/xml/9/34779/L.179.pc (accessed 28 December 2009).

Emmanuel, K.P. (2010) 'A Strategic framework for sustainable water resource management in a island economy with low freshwater availability: A case study of Barbados', unpublished PhD thesis, Environmental Management, University of the West Indies, Jamaica.

Gomes, A. (1993) 'Integrating tourism and agricultural development', In D. Gale and J. Goodrich (eds) *Marketing and Management in the Caribbean*, London: Routledge, pp. 155–66.

Gooding, E.G.B. (1971) 'Food Production in Barbados with particular reference to tourism' in G. Doxey *et al.* (eds) *The Tourist Industry in Barbados*, Ontario: Dusco Graphics, pp. 74–116.

Goodman, D. (2003) 'The quality 'turn' and alternative food practices: reflections and agenda', *Journal of Rural Studies*, 19: 1–7.

Goodman, M. (2004) 'Reading fair trade: political ecological imaginary and the moral economy of fair trade foods', *Political Geography*, 23(7):891–915.

Goodman, D. and DuPuis, E.M. (2002) 'Knowing food and growing food: Beyond the production-consumption debate in the sociology of agriculture', *Sociologia Ruralis*, 42(1): 6–23.

Harris, E.M. (2010) 'Eat local? Constructions of place in alternative food politics', *Geography Compass*, 4(4): 355–69.

Hinds, D. (2003) *Agro-Tourism Linkages Database*, Barbados: IICA.

IICA (2008) *National Strategy for the Promotion of Agri-Tourism Linkages in Barbados Agro-Trade and Culinary Dimensions*, Barbados: Agrico Ltd.

McKercher, B., Prideaux, B., Cheung, C. and Law, R. (2010) 'Achieving voluntary reductions in the carbon footprint of tourism and climate change', *Journal of Sustainable tourism*, 18(3): 297–317.

Marsden, T. and Smith, E. (2005) 'Ecological entrepreneurship: sustainable development in local communities through quality food production and local branding', *Geoforum*, 36: 440–51.

Momsen, J.H. (1986) 'Linkages between Tourism and Agriculture: Problems for the Smaller Caribbean Economies', Seminar Paper 45. Department of Geography, University of Newcastle upon Tyne.

Momsen, J.H. (1990) '*Survey of Barbados Hotels and Restaurants*', April. Geography Department, University of Newcastle upon Tyne. Unpublished.

Momsen, J.H. (1998) 'Caribbean tourism and agriculture: New linkages in the global era?', In T. Klak (ed.) *Globalization and NeoLiberalism: the Caribbean Context*, Lanham, MD: Rowman and Littlefield Publishers Inc., pp. 115–34.

Momsen, J.H. (2004) 'Post-colonial markets: New geographic spaces for tourism', In D.T Duval (ed.) *Tourism in the Caribbean. Trends, development, prospects.* London: Routledge, pp. 273–286.

Momsen, J.H. (2005) 'Caribbean peasantry revisited', *Southeastern Geographer*, 45(2): 206–21.

Momsen, J.H. (2006) 'Sustainable food for sustainable tourism in the Caribbean', In J. Hill, A. Terry and W. Woodland (eds) *Sustainable Development: National Aspirations, Local Implementation*, Aldershot: Ashgate, pp. 159–74.

Momsen, J.D. and Aspinall, R.J. (1987) 'Small Scale Agriculture in Barbados, 1987', Paper No. 52, Department of Geography, University of Newcastle upon Tyne.

Patullo, P. (1996) *Last Resorts: The Cost of Tourism in the Caribbean*, New York: Monthly Review Press.

Rhiney, K. (2008) 'Towards a new model for improved tourism-agriculture linkages? The case of two farmer's cooperatives in Jamaica', *Caribbean Geography*, 15(2): 142–59.

Richardson, P.E. (2004) 'Tourism and agriculture in Barbados: understanding linkages', Unpublished Undergraduate Thesis, Department of Geography, University of Edinburgh.

Richardson, P.E. and Whatmore, S.J. (2009) 'Food networks', In R. Kitchin and N. Thrift (eds) *International Encyclopedia of Human Geography*, Oxford: Elsevier, pp. 202–07.

Richardson-Ngwenya, P.E. (2009) 'The Ethical Geographies of Caribbean Sugar', Unpublished DPhil Thesis, School of Geography and the Environment, University of Oxford.

Richardson-Ngwenya, P.E. (2010) 'The EU Sugar Reform and the responses of Caribbean sugar producers', *Geography*, Vol. 95 Part 2: 70–9.

Sergent, J. (1967) *Report of the Tripartite Economic Survey of the Eastern Caribbean*, London: Ministry of Overseas Development, HMSO.

Thangaraj, P. (2009) 'Tourism sector doing well', *The Barbados Advocate*, 10: 7 and 21 October.

Torres, R.M. and Momsen, J.H. (2004) 'Challenges and potential for linking tourism and agriculture to achieve pro-poor tourism objectives', *Progress in Development Studies*, 4: 294–31.

Wilk, R.R. (1999) '"Real Belizean food": Building local identity in the transnational Caribbean', *American Anthropologist*, 101(2):244–55.

World Travel and Tourism Council (WTTC) (2010) 'Caribbean – Key Facts at a Glance'. Online. Available at: www.wttc.org/eng

Part III

New forms of tourism and agriculture production and consumption

10 Adopting a sheep in Abruzzo: agritourism and the preservation of transhumance farming in central Italy

Rosie Cox, Lewis Holloway, Laura Venn, Moya Kneafsey and Elizabeth Dowler

Introduction

This chapter looks in detail at one farm in the Abruzzo region of central Italy which has used tourism as part of an innovative mix of value-adding initiatives to re-establish traditional transhumance sheep rearing in a rapidly depopulating and economically marginalized area. The Italian government encourages and supports agritourism ('agriturismo') projects, in which an explicit link is made between farms, their produce and the accommodation and catering available, as part of its rural development strategy (Sonnino 2004). This scheme is therefore part of a wider trend within Italian rural spaces as well as being unique in the specific way it has used agritourism to support rural development. The case-study farm offers traditional agritourism on site and has developed a 'village hotel' using empty housing as permanent residents have left. It also has a shop, a restaurant and a classroom and hosts student groups. This complex of activities is underpinned by an Internet-based scheme, 'Adopt-a-Sheep' (see www.laportadeiparchi.it/help_en.htm), which allows people from all over the world to 'adopt' a milking ewe and receive farm products (cheese, salami and woolly socks) in exchange for an annual fee. The scheme, which has over a thousand members, guarantees an income to the farm and supports the agritourism enterprise as 'parents' are encouraged to visit 'their' sheep. Thus, the successful maintenance of a traditional mode of farming, rooted in a very particular place, is supported by global connections and made possible by high technology.

In this chapter, we trace the network of activities associated with the Adopt-a-Sheep scheme and La Porta Dei Parchi farm where it is based. The chapter describes the relationships between agritourism, food production and farming methods and the Adopt-a-Sheep scheme. It shows how these activities have been developed as an interlocking chain to increase the value added to agricultural products. However, rather than this arising from a farmer's desire to increase profitability at all costs, all the activities at Porta dei Parchi spring from a thought-through, deliberate initiative to maintain a particular environment and associated ways of life. Although many agritourist initiatives are part of farm diversification plans developed to increase farm incomes and to maintain farm businesses, in the case of La Porta dei Parchi farm, farming itself and all the other associated activities exist to preserve a particular rural environment. Agritourism is important to this both as an income stream and as a way to engage people in the wider project by invoking their enthusiasm for the landscape and traditional modes of farming intrinsic to it. Additionally, the produce made on the farm, which is sold on site and globally through the Adopt-a-Sheep scheme, is marketed in a way that links produce characteristics to the landscape and thereby valorises both the produce and the place from which it comes.

Research context

This case study formed part of a larger research project[1] exploring the restructuring of relationships between producers, consumers and food through participation in the so-called alternative food networks (AFNs; see Kneafsey *et al.* 2008 for more details on the project). Simple definitions of AFN are difficult, as the term has been used highly variably in accordance with differing agendas. Stassart and Whatmore (2003) write that, 'alternative production and consumption practices galvanise new modes of sense-making against the market and state-sanctioned rationalities of industrialisation' (p. 449). As such, AFNs tend to be associated with both spatialities and ethical frameworks that differ from those of 'conventional' food supply systems (Holloway and Kneafsey, 2004). Goodman (2003) has argued that the concept of AFN encompasses both those enterprises that can fit into the interstices or around the margins of the 'mainstream' industrial food supply system and more radical projects that are oppositional to industrial food supply and relate to a wider sense of protest and to attempts to establish different modes of exchange between food producers and consumers (e.g. Allen *et al.* 2003). The case of the Adopt-a-Sheep scheme and La Porta dei Parchi farm combines elements of both these approaches (Holloway *et al.* 2006).

Focusing especially on consumers' engagement with particular AFN projects, and their changing perception and understanding of food, food production and producers, our project involved an intensive research process that worked with six case studies, from very small-scale community garden projects to larger scale organic box delivery schemes and Community Supported Agriculture projects. Therefore, the project was not investigating the relationship between food and tourism directly but rather had traced the different ways in which food consumption fits into people's lives, be it as routine or exception. A particular focus of the research has been the ethical relations surrounding such AFNs, particularly the relations between food producers and consumers. For each of the UK-based projects, we carried out two rounds of in-depth, qualitative interviews with producers and consumers, focus group discussion with consumers and an ethnographically based 'household study'. As the case of the Adopt-a-Sheep scheme involved consumers from all over the world, we took a different approach. Questionnaires were sent to all adopters in English-speaking countries and in Italy (the questionnaire and responses were translated), an online discussion was hosted and field visits were made to the farm and interviews undertaken with the farm manager in 2004 and 2005.

The scheme discussed in this chapter is an unusual case that combines Internet-based direct sales with on-farm tourism, retail and education. Additionally, Adopt-a-Sheep and the other activities at La Porta dei Parchi have been developed deliberately to meet specific environmental and community-supporting goals. The desire to keep native sheep on the mountains to protect the environment and to provide employment for local people shaped all the activities on the farm. The ways in which agritourism activities developed in response to these aims is discussed in detail in the following sections. The chapter begins by giving the context for the development of agritourism in Italy and then goes on to elucidate how activities at La Porta dei Parchi have been developed to meet particular goals of protecting a specific landscape and associated ways of life. It looks first at the provision of accommodation on the farm and in the local village and then at the relationship between agritourism and speciality foods produced on the farm.

Agritourism in Italy

Agritourism is popular and widespread in Italy, with many hundreds of working farms supplementing their incomes by offering visitors the chance to stay in beautiful and sometimes remote places. Although it is now booming in Italy, agritourism developed relatively late here in comparison with other Western European countries. The Italian law enabling agritourism (la lagge quadro, 5 dicembre 1985 n.730) was passed in 1985, and by 1994, there were still only 7,000 agritourist businesses in the country. However, over the past 20 years, both agritourism and rural tourism more generally have been developing and are often portrayed as the solution to all the problems of rural areas (Ohe and Ciani 1998). Agritourist activities range from providing snacks for visitors who make brief visits to whole 'farm holidays', including accommodation, food and activities.

Agritourism in Italy is supported by government initiatives and also regulated by the government, and this influences the form it takes and the size of enterprises. Italian government policies to promote agritourism have been seen as a way to help revitalize rural areas by aiding the survival of agriculture in marginal areas, increasing farm income, creating new jobs in rural communities, adding value to regional agricultural products and diversifying tourism opportunities (Clemens 2004: 11). A key element of regulation is that farming activities must remain the primary source of income for a farm to be authorized to provide 'agritourism'. Italian farmers can be licensed to provide three different types of tourist service: snacks, full meals or meals and accommodation. In addition, the farm owner or operator must have 2 years of farming experience and, if he or she wishes to be eligible for funding from government programmes, must undergo 100 hours of training and pass an oral exam. The training programme covers law, farm management, accounting, hygiene, processing and transporting agricultural products and hospitality (ibid.).

Despite this support, the development of agritourism is uneven and certain areas have seen much greater growth than others. The most successful areas are broadly areas where the natural environment was already seen as a tourist attraction, close to the main urban cultural centres, with a long history of attracting tourists, sometimes over centuries, and those that are identified with typical food products. Businesses also reflect regional differences in culture and agriculture. For example, winter farmhouse accommodation for skiers in the Italian Alps was one of the earliest agritourism developments in Italy, whereas agritourism in Tuscany is promoted as a means for preserving rural cultural assets and buildings (Ohe and Ciani 1998).

Areas that might be thought of as more paradigmatic in terms of the Italian rural experience – characterized by remoteness, low incomes and low population density – are less likely to have seen the benefits of rural tourism (Marandola *et al.* 2006). However, agritourism is also expected to help develop the economy of areas that are characterized by depopulation, especially in southern Italy, while at the same time valorizing conservation (ibid.). Clemens (2004) comments that agritourism is also favoured as a method for keeping families on the land. Typically, farming families in marginal areas are extended and comprise more than one household, such as parents and their married children or married siblings and their children. Agritourism can be a means to generate enough additional income to prevent family members from leaving and thereby helps keep a workforce available for agriculture.

Agritourism is also regulated by regional laws and there are great differences in the types and number of agritourism enterprises in different regions. Regional regulations can affect the number of agritourism businesses in different areas by limiting the number of beds each farm is allowed to provide and the amount of income a farm can earn from tourism while still being recognized as an 'agritourism' business, rather than just 'rural tourism'. This distinction is important because agritourism businesses have greater recourse to government financial support. Agritourism organizations are active in differentiating 'agritourism' from 'rural tourism' by emphasizing their role in conservation and so justifying their requests for public support (Ohe and Ciani 1998).

Ohe and Ciani (1998) suggest that the central area of Italy has high rates of agritourism, with regulations permitting the largest number of beds per enterprise and allowing tourism to account for a much larger proportion of farm business. Farms in the central region are also more likely to provide activities such as horse riding to guests and to sell them farm products. However, within this large region there are also great variations. The Abruzzo region, where the Adopt-a-Sheep scheme is based, is characterized by depopulation, inaccessible mountainous terrain and low levels of agritourism. For example, the agriturismo.it website (see http://en.agriturismo.it for the English language version of the site) lists just 29 agritourism business in Abruzzo but 146 in neighbouring Umbria and 430 in traditionally popular Tuscany. Abruzzo is therefore more similar to southern Italy in some ways than to its central neighbours.

Adopt-a-Sheep and 'La Porta dei Parchi' farm

From here, the chapter focuses in detail on the case study of the Adopt-a-Sheep scheme and the farm and cooperative to which it belongs. The Adopt-a-Sheep scheme is based at a large mountain farm of 1,100 ha, the main enterprise being about 1,400 organic milking ewes. The farm is called 'La Porta Dei Parchi' (The Gateway to the Parks), reflecting its close proximity to two National Parks, and is part of a cooperative of 140 farms and other enterprises in the area. The case study farm is the largest farm in the cooperative and is by far the most dynamic in terms of both its range of diversification and in stimulating and managing a cooperative ethos.

The farm consists of semi-natural land located in a spectacular mountain landscape, which is designated as a National Park – a designation which suggests that sustaining the aesthetics of the landscape might be as important as ecological or economic sustainability. The terrain is rough, with harsh winters and predatory species (bears and wolves, which sometimes kill sheep, requiring constant shepherding and protective sheep dogs). Grazing is found up to 1,600 m altitude and transhumance agriculture is practised. In summer, sheep are taken up to the high pastures along with a mobile dairy and hut for the shepherds (who remain with the sheep for many weeks), both designed and manufactured specially and transported up the mountain by tractor. For the rest of the year, the sheep graze daily at lower pastures, walking long distances, and during harsh weather, the sheep live in the barn (Holloway *et al.* 2006).

The farm is managed by Manuela Cozzi, a dynamic and dedicated conservationist whose aim is to preserve the traditional transhumance agriculture that she believes best suits the mountainous landscape of central Abruzzo, protecting a particular ecology, including wild plant and animal species as well as sheep predators. In fact, the full name of the adoption scheme is 'Adopt a sheep defend nature' – clearly linking sheep farming to the protection of this particular habitat. The front page of the English language version of the website explains the context of Adopt-a-Sheep in these terms:

> The awareness raising campaign that our organisation, based in Anversa degli Abruzzi in the heart of the Parchi d'Abruzzo, is conducting is aimed at involving the entire population on an important problem: the abandonment of the mountain habitat and environmental degradation. [...] Considering the continuous reduction in the numbers of those who, though heirs of the millennial knowledge of sheep rearing, that is part of the pathway of human civilisation are masters in the skill of managing domestic animals through the use of environmental resources without altering them, on the contrary conserving and enhancing them, we feel obliged to launch this initiative.
>
> (www.laportadeiparchi.it/help_en.htm, translation by website authors)

The scheme is therefore imagined as preserving both a natural environment and a way of life that is linked to it and intrinsically valuable.

The farm manager is not originally from the area but moved there after agricultural college because of her commitment to traditional sheep farming as a way of life and a means to maintain particular landscapes and ecosystems. She is passionate in her dedication to the natural environment in general and to the local environment in particular. To attempt to protect this environment in the face of depopulation and falling prices for unprocessed agricultural products, she has developed a range of interrelated ventures that all aim to increase the value added to produce and therefore maintain the profitability of hill farming. These ventures include on-site food processing (of cheese, salami and wool products), a farm shop and restaurant, accommodation on the farm and in a 'village hotel', a farm classroom and accommodation for students. Agritourism was initiated in 1988, and in 1994, the on-farm dairy, farm shop and 'adopt a sheep' activities began. In 1996, an on-farm abattoir was added, and in 2001, the restaurant was opened on the farm, serving farm-produced and locally sourced foods. Since 2004, additional accommodation has been built on farm (which now takes up to 30 people) and future plans include adding more value to sheep wool products by moving wool processing and clothing manufacturing to the farm (when we visited wool was being sent to Tuscany for processing into baby clothes, blankets and socks).

This complex of activities is underpinned by the Internet-based scheme, 'Adopt-a-Sheep', which allows people from all over the world to 'adopt' a milking ewe and receive farm products by post in exchange for an annual fee. The scheme was launched in 1996 in a blaze of publicity. By coincidence, the farm manager announced its launch at a conference where the Italian minister of agriculture was speaking. The minister's name translates as 'Mr Shepherd' and as a result, he was encouraged to adopt a sheep. This made a good news story and ensured coverage for the scheme in newspapers around the world. In its first 48 hours, the Adopt-a-Sheep website had 200,000 hits and received around 500 emails a day (Kneafsey *et al.* 2008). Although interest has diminished since the initial launch, the novelty of the scheme still ensures a large number of visitors to the website.

By the time of our first field visit in 2004, the scheme had attracted 1,130 adopters, including 505 in Italy, 355 in the rest of Europe and 270 outside Europe (including the United States, Canada, South America, Australia, Japan, Singapore and India). For € 190, adopters receive a certificate of adoption, an identity card and photo of 'their' sheep – who they may also name, 2 kilograms of organic pecorino cheese, 2 kilograms of organic juniper-smoked ricotta, 1 pair of woollen trekking socks, 5 litres of olive oil, 1 kilogram of organic *salamelle di tratturo* (sheep salami) or as a vegetarian alternative, a kilogram of *Caciocavallo* (a sheep milk cheese aged in a traditional manner on wooden slats). Cheaper packages are also available, which include a reduced quantity and range of products, while local adopters can include manure as part of their package.

The Adopt-a-Sheep scheme in itself makes little profit, but according to the farm manager, it is the basis for making all other activities on the farm viable and sustainable. The scheme is important to the farm's wider strategy in two ways. First, it guarantees income because adopters pay in advance for produce and may continue to adopt year after year. This allows the farm to make investments and plan ahead. The farm manager explained that the reliability of Adopt-a-Sheep income was more important than the size of the business. Second, and perhaps more importantly, Adopt-a-Sheep acts as a direct marketing mechanism, giving the farm immediate access to over a thousand potentially sympathetic and interested customers, and through this publicity, access to many thousands of others. Thus, it supports the agritourism enterprise and encourages visitors to an area off the tourist trail. Adoptive 'parents' are encouraged to visit to see 'their' sheep and the farm where it lives, attracting customers for the farm accommodation, farm shop and restaurant – all key areas in which value is added to farm produce.

Visiting Porta dei Parchi

Agritourism was first initiated at the Porta dei Parchi farm in 1988, based on on-farm accommodation, and has grown significantly into a varied mix of accommodations and activities. Accommodation is available in a small number of rooms at the farm and a larger number of self-catering units called the 'village hotel', consisting of converted buildings dotted around the nearby village of Anversa. The properties are available because of high rates of out migration that have considerably decreased the village population. The village hotel model was deliberately adopted rather than increasing on-farm accommodation because it supports local businesses, such as the baker and grocer's shop, which in turn means they are more likely to remain open and permanent residents will have access to these services, which makes it more likely that they will be able to stay in the village. There is a small campsite and dormitories on the farm for students who visit to learn about organic and traditional farming methods. In addition, a barn has been converted into a 'farm classroom' where college students staying at the farm and local school students, who regularly visit, can learn while on site.

Tourists are encouraged to participate in activities on the farm as part of their trip. The activities provided all aim at helping visitors to understand and appreciate the local landscape and traditional methods of sheep farming. Tourists are invited to spend time with the sheep and the shepherds walking to new pastures, milking the sheep by hand or making cheese using traditional methods. During the summer, visitors can organize to go on horseback to visit sheep in the high pastures, and nature trails have been developed in the valley for those interested in less strenuous activity. These activities are a departure from more conventional agritourism ventures, which might serve farm products, but are unlikely to invite visitors to help make them. Providing accommodation on the farm is important in two ways. First, guests are a 'captive audience' for information about the farming methods and the local environment, linking to the broader project to preserve this particular way of life. Second, the accommodation provides some income, but more importantly, resident visitors provide a market for specialty products sold through the restaurant and shop.

Accommodation on the farm is advertised independently of the Adopt-a-Sheep scheme as well as in conjunction with it, but the scheme is understood to provide particularly valuable links with potential visitors. The Adopt-a-Sheep scheme works as a marketing tool not only because it gives the farm contact with 'parents' who might want to visit 'their' sheep but also because the novelty of the scheme attracts visitors to the website. The website includes pictures of the farm and surrounding landscape and text that gives some detail about traditional modes of farming and food production and their relationship to conservation. This marketing attempts three things. First, it draws attention to the sensual qualities of the local environment and farming practices, evoking a vicarious experience of tastes, smells, sounds and textures. In doing so, it engages viewers with the idea of the place as special but furthermore vulnerable and worthy of protection. Second, it makes a clear argument about the value of traditional rural ways of life, communities and environments and encourages the viewer to support these through their participation in the scheme and the farm. Last, it appeals to the viewer as both a connoisseur of quality foods and experiences and as an ethically engaged individual, thereby offering the viewer a positive identity to align oneself with if he or she chooses to adopt a sheep or visit the farm.

Our survey of adoptive parents suggests that people joined the scheme because of a desire to support rural farming. Members enjoyed the sense of belonging to a unique form of food production that operates on a small scale and protects both the physical environment and animals. When describing the positive aspects of the Adopt-a-Sheep scheme, most respondents cited respect for the environment and support of 'traditional' farming methods. Organic farming on a small scale and an ideal source for quality products were also consistently identified as benefits of the scheme. Only a minority of adopters do in fact visit the farm, but many more express the desire to visit and Manuela Cozzi told stories of individuals and families who return year after year.

Specialist food products

The quality and regional character of the products marketed through the Adopt-a-Sheep scheme are important elements in the success of the program and part of its link to agritourism. The strong link between place and products produced through the Adopt-a-Sheep scheme makes the appeal of both products and location stronger. The products are strongly portrayed as originating from a real place, made by real people who the consumer could meet and see at work. The place is valorized because of the quality and uniqueness of its 'traditional' and 'typical' produce. The mutual development of agritourism and regional produce has been noted as a trend in Italy, with 'typical' produce attracting visitors to farms. At La Porta dei Parchi, this trend has the added element of linking produce and location into a direct marketing scheme with global reach.

Many Italian farmers have discovered that farm tourists are prepared to spend relatively large amounts of money on eating and drinking on farm and purchasing local products. Farmers have thus embraced agritourism as a way to add value to their produce and market products made on the farm. Although conventional modes of farming involve transporting produce off the farm for processing and sale, agritourism allows the customer to both buy and sometimes consume the product on farm. In Italy, visitors regularly purchase wines, meats, cheeses, jellies and jams, honey and baked goods to take home (Clemens 2004). Ohe and Ciani (1998) found that over 80 per cent of agritourism businesses also sell farm produce to their guests. Direct sales have therefore become an important income source for farmers and an important part of the 'mix' for agritourism businesses. In addition to providing extra income, the sale of farm produce is one of the ways in which farms create a unique identity and differentiate themselves from more mainstream rural tourism.

Providing meals can also be a good way to retail farm produce to tourists. Ohe and Ciani (1998) found that about 70 per cent of Italian agritourism business offer lunch or dinner to guests, reflecting the Italian custom to eat at your hotel rather than paying only for a room or bed and breakfast. Typically, guests are not given a menu to choose from but are served the food that is available at the time. This is a common practice in small restaurants in Italy and allows farms to use their own produce as much as possible. In addition, Italian farmers are also permitted to make their own wine to serve guests and this is common practice at agritourism restaurants.

The Porta dei Parchi farm has followed these trends and developed a restaurant and farm shop as well as means of processing foods on site to add value to the milk and meat that is produced. The history of the business illustrates a continuous process of diversifying and adding value to the farm's basic product, sheep milk.

The farm manager made clear to us that her aim is to add value to produce whenever possible. For example, she explained that 5 litres of sheep milk was worth € 5 to the farm business. Processed into cheese to be sold by a wholesaler, the same 5 litres was worth € 12–15; yet, if sold directly to customers through a farm shop, it would be worth € 35. Further value could be added to the cheese-manufacturing activities through associated tourism (i.e. by people coming to look at what they do) and by serving the cheese at the restaurant. For Porta dei Parchi farm, the sustainability of the business enterprise relies on an intertwining of food production and tourism. The farm shop and restaurant were both important sources of income, but the restaurant was described as the pinnacle of the value-added chain as absolutely nothing leaves the farm. Products are produced, processed and consumed in place.

Agritourism policies in Italy complement policies encouraging production of local products by requiring that a minimum percentage of the food served to tourists is produced on the farm and in the region (Clemens 2004) which in turn supports the production of regional speciality food and drink products. Producers can add value to their products by applying for legal recognition of the direct connection between the high quality of their products and the specific place of production. This can be achieved either through national legislation or the European Union's scheme that endows food and drink with either a 'protected designation of origin' or a 'protected geographical indication' (PGI). Such 'appellations' enable producers to charge a premium for products that are thus recognized as being of specified geographical origin, high quality and often produced using traditional, artisanal methods (see, e.g. Ilbery *et al.* 2000; Parrott *et al.* 2002 on PGIs and Bessière 1998; De Roest and Menghi 2000; Tregear *et al.* 2007 on links between quality or local produce and rural development).

Clemens (2004) argues that in Italy, a number of trends in agriculture have come together to make agritourism and speciality food production mutually beneficial. For example, Italian food policy now encourages the production of high-quality, high-value products rather than only high-volume commodities. Additionally, there is an emphasis on adding value to agriculture through agritourism and this has great appeal in Italy where many traditional cultural activities are related to eating and sharing food. At the same time, EU regulations have protected speciality food producers from competition. Together these trends have supported both agritourism and speciality food production.

The rising popularity of traditional products is also a reaction to consumer demand for lower pesticide, herbicide and fertilizer use. By returning to traditional crops that expand diversity and using historical production methods, producers are in some respects responding to demands of the organic movement.

The Porta dei Parchi farm produces high-quality, high-value 'typical' produce in the form of cheeses (which have won medals at the 'Mountain Cheese Olympics') salami, preserved fruits and liquors, all of which are certified organic. The majority of this produce is sold relatively locally and some are sent regularly to shops and restaurants throughout Italy. For this produce, recognition as regionally typical helps the farm earn a better price. However, in concert with the Adopt-a-Sheep scheme and agritourism at the farm, the recognized 'typical' nature of these products takes on increased value. The strong associations between the farm, its particular farming methods and the environment it is located within, present in marketing literature for Adopt-a-Sheep and agritourism at the farm, invite visitors and adopters to understand the products they are eating or buying as intrinsically part of a particular landscape and environment. The message that some products are unique to and beneficial for certain places is particularly strong in this context and serves to increase the value of both the products and the place.

Conclusion

In many ways, this chapter recounts what appears to be a success story in finding innovative ways to integrate agritourism with traditional modes of production and environmental protection by reaching out to markets well beyond the local or national space. La Porta dei Parchi farm has developed an interrelated group of activities: sheep farming, food processing, retailing through the shop and restaurant, accommodation and direct sales, which reinforce each other to maximize the value added to produce without threatening environmental or social sustainability.

The use of an Internet-based marketing scheme combined with a strong and consistent conservation message make the farm particularly unusual. The Adopt-a-Sheep scheme provides a reliable source of income and acts as a marketing and outreach tool allowing the farm manager to communicate her goals globally while simultaneously underpinning the activities (food processing and retail) that allow transhumance farming to be sustained. Thus, the successful maintenance of a traditional mode of farming, rooted in a very particular place, is supported by global connections and made possible by the high technology of the Internet.

The goals of La Porta dei Parchi farm and the Adopt-a-Sheep scheme were preserving a particular type of rural economy, community and environment characterized by 'traditional' lifestyles, foods, relations of production and human–livestock–nature relationships. These goals attribute value to a specific local environment and to food produced there in certain ways. The farm's message valorizes both its products and the place they come from, in turn encouraging both agritourism and food sales, on farm and at a distance. As a result of these strong associations, the produce markets the place and the place markets the products – even to consumers on the other side of the world.

Note

1 The research project was made possible by funding from the ESRC and AHRC (Grant number RES-143-25-005) as part of the Cultures of Consumption research programme.

References

Allen, P., Fitzsimmons, M., Goodman, M. and Warner, K. (2003) 'Shifting plates in the agrifood landscape: the tectonics of alternative agrifood initiatives in California', *Journal of Rural Studies*, 19: 61–75.

Bessière, J. (1998) 'Local development and heritage: traditional food and cuisine as tourist attractions in rural areas', *Sociologia Ruralis*, 8: 21–34.

Clemens, R. (2004) 'Keeping farmers on the land: Adding value in agriculture in the Veneto region of Italy', MATRIC Briefing Paper 04-MBP 8, Midwest Agribusiness Trade Research and Information Center.

De Roest, K. and Menghi, A. (2000) 'Reconsidering "traditional" food: the case of Parmigiano Reggiano cheese', *Sociologia Ruralis*, 40: 439–51.

Goodman, D. (2003) 'Editorial: the quality "turn" and alternative food practices: reflections and agenda', *Journal of Rural Studies*, 19: 1–7.

Holloway, L. and Kneafsey, M. (2004) 'Producing-consuming food: closeness, connectedness and rurality', in L. Holloway and M. Kneafsey (eds) *Geographies of Rural Cultures and Societies*, Aldershot: Ashgate, pp. 262–82.

Holloway, L., Cox, R., Venn, L., Kneafsey, M., Dowler, E. and Tuomainen, H. (2006) 'Managing sustainable farmed landscape through 'alternative food networks' a case study from Italy', *Geographical Journal*, 172: 219–29.

Ilbery, B., Kneafsey, M. and Bamford, M. (2000) 'Protecting and Promoting Regional Speciality Food and Drink Products in the UK: the Case of PDOs and PGIs', *Outlook on Agriculture*, 29: 31–7.

Kneafsey, M., Cox, R., Holloway, L., Dowler, E., Venn, L. and Tuomainen, H. (2008) *Reconnecting Consumers, Producers and Food: Exploring Alternatives*, Oxford: Berg.

Marandola, D., Cannata, F., De Palma, P., Fragnito, S., Parletta, I. and Raschi, A. (2006) 'Business and tourism in a rural area of southern Italy', *Geotour*, October 2005: 157–61.

Ohe, Y. and Ciani, A. (1998) 'The activity characteristics of agri-tourism farms: a study of Umbria,' *Bulletin of Chugoku National Agricultural Experimental Station*, 19: 1–18.

Parrott, N., Wilson, N. and Murdoch, J. (2002) 'Spatializing quality: regional protection and the alternative geography of food', *European Urban and Regional Studies*, 9: 241–61.

Sonnino, R. (2004) 'For a ''piece of bread''? Interpreting sustainable development through Agritourism in southern Tuscany', *Sociologia Ruralis*, 44: 285–300.

Stassart, P. and Whatmore, S. (2003) 'Metabolising beef: food scares and the un/remaking of Belgian beef', *Environment and Planning A*, 35: 449–62.

Tregear, A., Arfini, F., Belletti, G., Marescotti, A. (2007) 'Regional foods and regional development: The role of product qualification', *Journal of Rural Studies*, 23: 12–22.

11 Farm-stay tourism in California: the influence of type of farming

Jill Donaldson and Janet Momsen

There is a long tradition of farm-stay tourism in Europe but in the United States, it only reached California at the end of the twentieth century after becoming established in New England a little earlier. Farm-stay tourism is a subset of rural tourism or agritourism in which visitors actually stay on farms. It has been defined as 'the housing of tourists in rural homes which continue to practice farming activities' (Canoves and Villarino 2002: 92). The growth of this type of tourism has been encouraged by the diversification of the tourism market from standard family groups to individuals travelling alone or in groups with similar interests, such as bird watching or groups of similar ages. In the United States, where paid vacations are much shorter than in Europe, increasing pressures to reduce distance travelled to minimize use of expensive petrol (gasoline) and to reduce carbon footprints have encouraged short stay vacations in nearby rural areas. In California, there is increasing pressure to protect farmland as population and housing spread out of the cities.

Rural tourism has seen the need to diversify its offerings in response to these consumer changes. Farm lodgings for tourists are no longer just thought of as places to stay in a rural environment but also as offering special pursuits to complement the stay. Such on-farm activities in California may include classes in cheese or jam making (Calaveras County), watercolour workshops (Alpine County), horseback rides (Fresno County and Kern County), gardening classes (Mendocino County), carriage rides (San Luis Obispo County), a fly fishing school (Siskiyou County), or gold panning (Trinity County) (Small Farm Center California Agri-tourism Database 2010).

In this chapter, we look at the beginnings of farm-stay tourism in California a decade ago and compare it to the situation today. This comparison is based on a survey of farm stays by Donaldson for her Master's thesis in 2000 (Donaldson 2002) and a survey in 2009 undertaken by the University of California at Davis (UCD) Small Farm Center of all agritourism enterprises (Leff 2009a). In 2008, Californian farmers and ranchers hosted more than 2.4 million tourists. The 2009 Agritourism Survey found that of these tourists, half came from the same county as the farm tourism operation, 38 per cent from other parts of California, 8 per cent from other parts of the United States and 3 per cent from foreign countries. This supports the idea that the growth in rural tourism is related to increased interest from local visitors.

Urbanization and shrinking profits have led Californian farmers seeking to diversify into new income-producing on-farm activities such as agritourism. The majority of these farms are small. In the 2009 survey, 68 per cent of respondents reported gross revenues of less than US$ 250,000, which means they met the US Department of Agriculture (USDA) definition of a small farm by revenue (Leff 2009a: 11). In 2007, the Census of Agriculture reported that 76 per cent of Californian farms were considered small farms as defined by gross revenue (USDA 2009: 64). Between 2002 and 2007, the US Census of Agriculture recorded that the total number of farms in California fell from 81,033 to 79,631 (USDA 2009: 17). The only size of farm to show an increase over this period was the class of less than 50 acres (20 hectares), which increased from 49,134 farms in 2002 to 53,368 in 2007 (ibid.). The number of farms in this class had remained virtually static from 1982 to 1987 reaching 51,195 in 1987 (US Bureau of the Census 1989: 1). Clearly, the new millennium showed a turnaround in the growth of small farms in California. This may be linked to the growth of pluriactivity, especially tourism. Almost half (47 per cent) of the operators in the 2009 survey reported gross revenues from their agritourism activities of less than US$ 10,000, but 22 per cent reported gross agritourism revenues of more than US$ 100,000 (Leff 2009a: 11). Thus, for many farms, income from agritourism is a vital part of their farm diversification and survival strategy.

A sabbatical visit from a Spanish agritourism specialist (Gemma Canoves, see Chapter 4) to the UCD in 1999, who queried the absence of these activities in California, shortly after Rilla (see Chapter 12), a member of Cooperative Extension at the UCD, had spent her sabbatical leave visiting farm stay operators in New England and Britain (Rilla 1997), triggered interest in farm-stay tourism in particular. The Director of the Small Farm Center at UCD had also just obtained funding to study statewide agritourism. This confluence of interest led to the setting up by the UCD Small Farm Center of the Agriculture and Nature Tourism Workgroup, a statewide group organizing research and education on agritourism in 1999. At this time, the California Travel and Tourism Commission considered rural tourism to be all tourism activities taking place outside the urban areas of San Francisco, Los Angeles and San Diego. The state Tourism Commission then sent a representative to the university-based workgroup and in 2002 published a *Directory of California Agri-Tourism* (California Tourism 2002) containing 364 locations offering various types of rural tourism activities. These activities mostly consisted of on-farm sales of various, mainly organic, products from meat to cheese and Christmas trees, farm and factory tours, museums and agricultural festivals celebrating a particular crop such as avocadoes in San Diego. There were only 11 places listed in this directory as offering farm stays.

The workgroup now has over 50 members including both academic and Cooperative Extension university members and many non-university collaborators from across the state. It was found that lack of state support and worries about legal liability discouraged the development of agritourism, especially farm-stay tourism in California. Gradually, these difficulties have been overcome largely through the organizing of grassroots farmer entrepreneurs and lobbying of state politicians to remove legal restrictions, facilitated by the UCD group. Since 1999 over 2000 Californian farm and ranch operators have attended Cooperative Extension-led workshops on how to diversify their businesses with agritourism. The workgroup also produced a handbook, *Agritourism and Nature Tourism in California*. This has sold over 8,000 copies since it was first published by the university in 2002, and it is now in its second printing. With assistance from the workgroup, state support of legal changes has eased the problem of legal liability for accidents to farm visitors and technical requirements for serving meals to guests and running roadside stands selling farm produce and thus has encouraged a growing number of farmers to set up various on-farm activities for visitors including farm stays (Leff 2009b: 10). However, the 2009 survey conducted by UCD revealed that of the respondents, 25 per cent had begun their agritourism operations before 1984 and only 43 per cent since 2000. For those early agritourism innovators, farm stays were not common, although some advertised Bed and Breakfast facilities. The most common agritourism activity was probably direct sales of agricultural products, which in 2009 provided 45 per cent of all agritourism gross income (Leff 2009a: 11). Farm stays and Bed and Breakfast offerings provided, on average, only 5.2 per cent of gross income directly (http://ucanr.org/agtour/prelim-results.pdf). Among the respondents in 2009, 15.1 per cent had farm stays, 8.8 had farm stays for which they charged a fee, 10.6 per cent offered camping facilities and 7.6 per cent had cabins for visitors (ibid.). Expanding or diversifying their operations over the next 5 years was planned by 65 per cent of respondents, whereas only 4 per cent planned to discontinue their agritourism activities (ibid.).

Survey of Californian farm-stay operations

In 1999, when the Small Farm Center Workgroup was set up, very little was known about agritourism in California. The motivation for the establishment of the workgroup had been a curiosity about the absence of farm stays in California when compared with other parts of the world. Today the Agri-tourism Database lists 97 farm-stay operations in California out of a total of 706 sites, so they are still a minority type of agritourism activity (www.calagtour.org/list.asp, accessed 25 March 2010). They were found in 39 counties of a total of 58, all over the state, but most were concentrated in San Luis Obispo County and Tulare County (9 each), San Diego County (7), Sonoma County (6) and El Dorado County (5). In terms of the 12 official tourist regions, all have farm stays: most were on the central coast with 17, followed by the north coast with 13, the south Central Valley with 12 and the foothills known as the Gold Country where there are many small wineries and olive oil producers with 11 farm stays (ibid.).

When we started to look for farm-stay operations in 1999, we took the entire state of California as our research site (Donaldson 2002). As farm-stay tourism was only just beginning in the state, the search for examples was a challenge. We first contacted farmers who had attended the early workshops of the Agritourism Workgroup. From their initial responses, we followed a snowball technique.

We also searched Internet sites and the California Association of Bed and Breakfast Inns website and eventually found a total of 17 farms offering farm stays. These were in the Gold Country, the North and Central Coast regions (Napa Valley, Marin and Paso Robles), the mid-Central Valley and the extreme north of the state. Almost three quarters of the sample were located in or near the Gold Country, Paso Robles or the Central Valley. At this time, we did not hear of any farm-stay operations in the southern part of California, probably because we were working from Davis in the North Central Valley region and the workgroup's early contacts were in relatively close proximity to the university campus. Nevertheless, only the South Central Valley and the South Coast with major concentrations of 12 and 8 farm-stay operations, respectively, listed in 2010 were not represented. We included all 17 farms that fitted the working definition of a farm stay in our research study. A total of 15 of the respondents stated that their area was popular with tourists but this was a highly subjective response. At least four of the respondents were located so far off the beaten track that tourists had to make a special effort to seek them out. In total, 15 of the 17 farm stays studied had been founded in the previous 10 years. This is in stark contrast to the age of the farms themselves: five had been founded in the nineteenth century and two more in the early 1900s, but most farms studied had been in existence for less than 20 years at the time of the survey. About half the sample had planned a joint agriculture/tourism venture from the beginning, whereas others had revamped their farm operations with a new reliance on both the food product and the farm stay.

The sample group consisted of five ranches (cattle and timber), five farms (apples, almonds, lavender, honey, ostrich and dairy cattle) and five wineries (vineyards and wine makers) and two mixed operations. The mixed use farms, one in Amador County and one in Marin, were both wineries but one also had a dairy while the other had beef cattle and sheep, walnuts and flowers. One third of the sample enterprises incorporated organic practices, but this was not used as a selling point to attract tourists. There was a wide range of type of accommodations, price range, activities and atmosphere because each of the farms had their own unique attraction and target market. Wine tourism has been seen as the godfather of agritourism in California and is highly developed and very popular. It is very different from farm-stay tourism, but the farm stays we identified as linked to wine producers were generally outside the main wine areas and fit the study definition. The survey instrument was tested on workshop participants and then revised with reference to one used in Australia by Fry (1984). Eventually, the survey was administered in 2000. Content analysis of brochures and promotional material and information from photographs and research notes were also used.

The activities offered at the farms studied ranged from hiking/walking (15), bird watching (12), farm tours (11), cycling (10), harvest festivals (8) and swimming (7). One farm did not answer this question, but clearly the most common offerings were outdoor recreation rather than educational. The unique features that some of the farm-stay operators provided to lure guests were varied and often played a major role in the success of the venture with the cookery classes at the Apple Farm, belonging to a former owner of the iconic French Laundry restaurant, being booked up well in advance. Other offerings included wine tasting, cattle drives, guided hunts, beekeeping and weddings.

Characteristics of farm-stay surroundings

All the farms in the study fell into one or more of three categories. Generally, each of these categories catered to a particular type of tourist, with wine tourism being the most established of the three. The environment of each operation was usually linked to the group to which it belonged. For example, all ranches were characterized by large sweeping landscapes and a long-standing rural tradition, wineries had nearby vineyards and some sort of gift shop or tasting room and farms concentrated more on the agricultural activities of the property and nearby attractions. Additionally, almost all proprietors offered some unique feature that served as the prime attraction for the visitor. Viewed as atmosphere or specialized activities, these features allowed the enterprise to appeal to a certain type of person, which led to fewer problems with guests and assisted in both repeat business and word of mouth advertising.

On the basis of visits, photographs and promotional brochures, 15 of the 17 farm stays sampled were judged on a 5-point scale in terms of their surrounding scenery, gardens and accommodation. A total of 11 were judged excellent for their scenic landscape and for well-kept farm buildings. Six were judged good in terms of the modernity and comfort of their guest rooms, whereas five were thought to be excellent in this category and six were considered average for room decor. Combining all scores for room quality, six were considered excellent and had high occupancy rates and strong repeat business.

Differentiation of farm-stay type proved important in other ways as well. Each of the three categories showed patterns related to specific goals sought from tourism. Wine-related enterprises were concerned with using tourism as a marketing tool to help sell their product. Farms concentrated on maximizing the additional income provided by overnight guests and the promotion of local products in the hope that this would both raise product demand and educate the public on the importance of small farms to the state's rural areas. Ranchers too were mindful of the need to educate the public, the main goal being to make visitors, usually from urban areas, aware of the value of preservation of large land holdings both ecologically and aesthetically.

Most of the older farms emphasized historical aspects of California, including the architecture and the importance of rural traditions. Most farms aimed to attract particular types of tourists as many guests learned of the farm stay through word of mouth recommendations. The specific farm product was of major importance in attracting tourists for eight farms surveyed of which six were wineries. Education about these products was seen as an attraction by five farms. These findings indicate the importance of marketing farm products to guests in addition to simply the rural experience. The most important goals for these proprietors of farm stays were additional income followed closely by diversification. These motivations varied by type of farm (Table 11.1).

Table 11.1 Motivations for starting a farm-stay business

Total scores for eight motivations	
Additional income	77
Diversification	69
Enhancement of sustainable farming	62
Preservation of farmland	61
Educate the public about rural life	60
Raise product demand	58
Creation of jobs	47
Promotion of local farms	46

Source: Adapted from Donaldson (2002: 87).

In the 2009 Agritourism Survey, the main motivator for farm operators deciding to open their farms to visitors was for 75 per cent of the farms to increase profitability and secondly to educate visitors (64 per cent) followed by use as a marketing tool (62 per cent). This survey was focused on general agritourism, not just on farm stays, but the motivation was very similar to that found in the earlier survey. For advertising, most operators in both surveys used word of mouth, websites and feature stories in local media and felt these were the most effective types of promotion.

Outreach through advertising was necessary for the farm stays to achieve their specialist goals. Methods of advertising often served multiple purposes giving them value far beyond simple exposure. Free methods were most popular, with word of mouth being most sought after and local networking coming second. These methods served to help ensure a high quality of guest and proved extremely effective in drawing clientele. The local networking had the additional benefit of strengthening community ties and reinforcing ideas of locality and regionalism among the tourist base. As similar farm stays were often clustered in certain localities, it meant that if one destination was full, tourists could be passed on to a neighbouring farm. Other networking opportunities presented themselves in the form of organizational membership, which also served as a forum for cooperative advertising. This was a cost-effective method of spreading information about farm stays. Finally, the Internet was considered by many to be an excellent way to increase information accessibility and offered the possibility for online mail order of farm products. The online list of farms offering agritourism opportunities maintained by the UCD Small Farm Center (www.calagtour.org/lsit.asp) has been a very popular statewide source.

Financial returns from agritourism

Four of the farm-stay operations in 2000 said they could not live without it financially. Of these, two were newer operations that had had overnight accommodations in their business plans from the beginning. The other two growers had diversified into farm stays after the agricultural enterprise was well established and this tourism was so successful that it had become the major focus of the business and critical to their survival. Six interviewees felt that the farm-stay side of the farm business significantly raised their standard of living and added to the financial security of the farm. Four felt that the farm stay was of minor importance. Of these, one was just starting and one was phasing out the farm stay, whereas the other two had not put much effort into the farm stay and were having trouble attracting visitors. The farmer who listed his farm stay as not important was just starting the business. Two interviewees chose not to answer this question.

Wine-related operations reported a slightly increased financial dependence on farm-stay income. These operations saw the overnight accommodations as a marketing tool to help move their wine as a direct source of income. Half of the farms also realized the potential in selling products to visitors. Some of the benefits of exposing guests to their products were directly financial, but the farmers also expressed a deep desire to promote local farm produce. In the general survey of agritourism in 2009, 45 per cent of gross agritourism revenues came from direct sales, whereas farm stays accounted for only 5 per cent. However, assuming relationships between sources of revenue were much the same as in the earlier survey, then farm stays could be seen as a way of marketing retail product sales both while visitors were on the farm and afterwards as a result of familiarity with the producers of certain products.

Labour utilization

The majority of the sample farm-stay operations depended on extended family labour. Much of the work could be seen as gendered (see Chapter 12). On most farms, women were in charge of the farm-stay operation, whereas men spent more time on agricultural operations. This could vary with the age of the farm operator. Where farms hired labour to deal with farm-stay work, it was usually part-time and seasonal. Mostly women were hired as the extra workers needed for cleaning of guest rooms, cooking and sales in gift shops. Canoves and Villaarino (2002) found a similar situation in Portugal, where members of the farm family were often working off the farm or had emigrated. Sometimes family members would be available to help at weekends, but most farm women found that guests increased their work hours (ibid.). Taking in tourists is seen as complementary to women's main job on the farm and not as a separate occupation (Garcia-Ramon *et al.* 1993). Studies in Europe often suggest that despite the limited economic gain, one of the benefits of farm-stay tourism is social, especially for isolated farm women (Kinnaird and Hall 1994). '[T]he social benefits of rural tourism and the enjoyment to be gained from this work may well outweigh economic limitations for most women' (Canoves and Villarino 2002: 100).

Problems and concerns

The initial hurdle to overcome when considering a farm stay was the dearth of readily available, relevant sources of information. Farm-stay operators felt they needed information on insurance coverage, labour issues, and health, building and zoning regulations. Many operators were not sure they had adequate insurance coverage. Labour issues usually involved problems of finding good help, legal and employment tax issues. Navigating the various regulations was a major difficulty as each county had their own. The Agritourism Workshop was a great help in providing training sessions and directing farmers to the most useful sources of information. In the 2009 survey, 29 per cent of respondents had acquired a use permit from their county for their agritourism operations. Over two thirds of them had found the permitting process expensive, difficult and slow (Leff 2009a: 11). Although 87 per cent of respondents were covered by liability insurance, most for US$ 1 million or more, they all said that insurance was a major challenge along with permitting, zoning and other legal constraints (ibid.). For the farm-stay operators interviewed in 2000 concerns included reactions from the surrounding community, financial returns, effects on farming and the family, privacy and type of clientele attracted. When neighbours were close and narrow roads or driveways shared, the increased traffic caused by visitors led to friction. Fortunately, this was not a problem in many areas. However, in some places, schedules for the use of noisy machinery and for crop spraying had to be organized to reduce the impact on paying guests. Interaction with guests also caused friction with family, especially when common space was shared and some farmers felt as if they were on duty all the time. Such friction was reduced when guest quarters were far removed from family areas. Most of these concerns lessened with time and experience.

Conclusion

This chapter has set out to establish the nature of farm-stay tourism in California and how it has changed over the last decade. California was very slow in developing farm-stay tourism when contrasted with Europe, Australia and New England. Zoning regulations and legal liabilities in California made it difficult for those wishing to start a farm stay to find the necessary information needed. Those who had planned their agricultural and tourist operations simultaneously found it easier as many of these proprietors came from a business background and were familiar with planning regulations. Owners operating within a long-standing agricultural enterprise had more difficulty complying with code and making the transition from cosy family homestead to allowing strangers as visitors into their personal space.

There was no well-known example to follow in California, so many farm stays modelled themselves on the familiar Bed and Breakfast pattern. This resulted in something of an identity problem in the farm-stay business as many of the farmers offering accommodation considered their farm stay to be a Bed and Breakfast Inn. Most farmers and tourists alike were unfamiliar with the farm-stay concept making it difficult to attract growth in its own right. However, we argue that this gave farmers the freedom to add descriptive and alluring terminology to their advertisements for their enterprises to attract the desired type of guest. As they were developing farm stays *de novo*, these Californian farms had a very special style. Today, when the term farm stay is well accepted, in Californian literature, these distinctions remain. Largely as a result of this organic development, farm stays in California are proving to be a good example of successful rural tourism.

Both Tourism and Community Development theory advocates a bottom-up approach to tourism planning to allow for the widest distribution of the benefits. When the drive for tourism comes from within and small-scale tourism is introduced gradually, it will be less disruptive, see a greater percentage of earnings retained locally and pave the way for a more harmonious relationship between the community, the provider and the tourist. Having substantial local input over a period of time allows the process to evolve so that the community may adapt the programme to their specific needs and limitations. This has allowed the farmers to retain their identity as a farmer (or rancher or wine maker) and has created additional benefits from that distinction. Furthermore, tourists are attracted to the small scale and individuality. So, while farm stays are in need of some outside assistance to create organization and momentum, they also need time to ensure a strong identity, a good reputation and a lasting presence. Fortunately, farm stays were able to establish themselves in California before the relatively recent increase in people seeking out meaningful tourism opportunities with a smaller carbon footprint.

These findings linking the nature of farm stays to the particular farm product appear to be unique to California, perhaps because the state has such a varied agricultural mix. These linkages were able to develop a decade ago as the farm-stay business started and can still be seen. The farm-stay industry still needs targeted outreach to the tourism industry and governmental assistance to overcome planning and legal restrictions. It has been shown that today when a few farms earn over US$ 1 million from agritourism, it has become a way of preserving farmland and keeping young people on the farm (see Chapters 10, 12 and 14).

References

California Tourism (2002) *California Agri-Tourism Directory*, Sacramento: A Joint Marketing Venture of the California Travel and Tourism Commission and The California Technology, Trade and Commerce Agency, Division of Tourism.

Canoves, G. and Villarino, M. (2002) 'Rural tourism, gender, and cultural conservation in Spain and Portugal', in M.B. Swain and J.H. Momsen (eds) *Gender/Tourism/Fun?* New York: Cognizant Communication Corporation.

Donaldson, J. (2002) 'An Assessment of Farm Stay Tourism in California', unpublished M.S. thesis, Community Development, University of California, Davis.

Fry, P. (1984) *Farm Tourism in Western Australia*, Perth: Rural and Allied Industries Council.

Garcia-Ramon, M.-D., Villarino, M., Baylina, M. and Canoves, G. (1993) 'Farm Women, Gender Relations and Household Strategies on the Coast of Galicia', *Geoforum*, 24(1): 5–18.

Kinnaird, V. and Hall, D. (1994) *Tourism: A Gender Analysis*, Chichester: Wiley.

Leff, P. (2009a) 'First statewide agritourism survey yields early results' *Small Farm News*, Vol. 2: 1 and 11.

Leff, P. (2009b) 'New farm stand regulations now in effect expand options', *Small Farm News*, Vol. 2: 1 and 10.

Rilla, E. (1997) *Unique Niches. A Report on a 1997 Sabbatical Study of Agritourism Development in Britain and Selected New England States*, Oakland: University of California, Agriculture and Natural Resources.

United States Department of Agriculture (USDA), National Agricultural Statistics Service (2009) *2007 Census of Agriculture-State Data*, California. Washington, DC: USDA.

University of California, Davis (2002) *Agritourism and Nature Tourism in California*, Davis, CA: Small Farm Center.

University of California, Davis (2009) Agritourism Survey. Online. Available at: http://ucanr.org/agtour/prelim-results.pdf (accessed 25 March 2010).

University of California Davis Small Farm Center (2010) California Agri-tourism Database. Online. Available at: www.calagtour.org/list.asp (accessed 25 March 2010).

U.S. (Department of Commerce) Bureau of the Census (1989) *1987 Census of Agriculture, Volume 1 Geographic Area Series, Part 5 California State and County Data*, Washington, DC: Bureau of the Census.

12 Tourism and agricultural viability: case studies from the United States and England

Ellen L. Rilla

Introduction

Agritourism in the United States has become popular in the past 10 years as one strategy for increasing and diversifying farm profitability (Brown and Reeder 2007). In England, the need for diversification and pluriactivity has long been recognized (Ilbery *et al.* 1998). Hilchey (1993: 2) defined the term agritourism as 'a business conducted by a farm operator for the enjoyment and education of the public, and to promote the products of the farm, and thereby generate additional farm income'.

In California, confusion over the definition led to reframing the term to include various historical, cultural and ecological assets a farm experience could provide (Donaldson and Momsen, Chapter 11). In this definition, agritourism includes any income-generating activity conducted on a working farm or ranch for the enjoyment and education of visitors. Agritourism includes the interpretation of the natural, cultural, historical and environmental assets of the land and the people working on it (George and Rilla 2008).

A majority of farms and ranches in the USA have diversified their operations to host some form of agritourism for economic reasons (Nickerson *et al.* 2001). Other research points to agritourism motivation being composed of a complex set of economic and social variables rather than any one predominant type of factor (Mace 2005). Ollenburg and Buckley (2007), in an Australian study, found that for most operators, both social and economic factors are important, and different motivations are dominant for different types of farm landholders and at different stages in farm, family and business lifecycles.

When asked about why farm operators decided to open their farms/ranches to visitors, a recent California survey (Leff 2009) also confirms the motivational mix of reasons behind the expansion of agritourism. Of the respondents (n = 332), 75 per cent wanted to increase the profitability of their farm business, whereas 69 per cent were also motivated by their desire to educate visitors about farming and ranching.

We know quite a bit about the motivational factors involved in farmers' decisions to initiate an agritourism enterprise from the studies previously cited. Other research reports have attempted to characterize the size and nature of farm agritourism in the United States (Brown and Reeder 2007) and in the United Kingdom (DEFRA 2004).

However, we do not know what has happened to those who have been operating agritourism venues for at least 10 years. Has it fulfilled their original goals? Do they plan to continue and why? How has it contributed to the viability of the farm operation?

My premise is that once begun, agritourism becomes so ingrained within the farm's economic, social and generational goals, it would be difficult, if not impossible to terminate. I am also interested in the interplay of economic, social and environmental goals and whether farming and tourism components are becoming less distinct over time. Although exploratory in nature, these brief case studies could lead to more comprehensive work on the viability of long-term agritourism operations and their role in sustaining farm families. Its findings are strengthened by the international comparison.

Methods

In 1997, I interviewed 100 operators on the west and east coasts of the United States and in the United Kingdom for an earlier project (Rilla 1998). Rather than re-interview all the original participants for this study, I did a simple web search in 2008 to ascertain what percentage had ceased operating. At that time, 90 per cent of the farms were still conducting some type of agritourism activity.

I then selected four operators I had originally interviewed in the three study areas to address the question of how agritourism venues influence long-term farm viability. Of particular interest is whether the agritourism enterprises continued, and possibly expanded, as families derived a mix of social, economic and community benefits that ultimately kept the farms viable. Farmers in Vermont and England were initially selected by park and governmental officials contacted in 1997 and referred to me. The two operators in California were the first in Marin County, north of San Francisco, to develop agritourism venues (I work in this County as extension faculty with the University of California at Davis).

Open-ended interviews were conducted with four women farmers involved in the three study areas between June and September 2009. The interviews lasted between 90 and 120 minutes and were audio recorded. The interviewees were informed at the beginning that the main focus would be on the endurance or staying power of their tourism endeavour as it related to keeping their farm viable.

I explored this topic through the following questions. Is the family still engaged in agritourism? Is the agritourism business still distinct from the farming operations? Have younger family members begun to manage operations? Is this an activity that they want to continue? Are there land use or governmental policies that have helped or hampered their efforts? And finally, what lessons can be learned that may be applied or adapted to other communities? Drawing upon this exploratory set of case studies, this chapter seeks to cast light on the sustainability of agritourism enterprises, as well as to inform future directions for agritourism research and policy. During the 1997 project, I observed that the majority of people who conduct some form of farm tourism – initiating it, developing it, forming associations and making it work – are women. Gender motivation regarding agritourism was previously surveyed in Virginia farm families (McGehee *et al.*

2007). The results of this study suggest that women are more motivated than men to undertake agritourism enterprises because of their consistently higher rankings for various factors such as family income, employee and community education. In addition, studies in Australia (Jennings and Stehlik 1999), the United Kingdom (Neate 1987) and Ireland (O'Connor 1995) all concur that agritourism is an innovation that has been taken up primarily by women partially due to cultural bias and assumptions about what constitutes 'women's work' such as management and visitor services. But also noted in some of the cited studies were the rewards and challenges a female farm partner derives in a traditionally patriarchal farm operation when initiating an agritourism business on the farm. Canoves and Villarino (2002) also noted a similar relationship in Spain.

In each of the three study areas, operators face high land costs and strict environmental regulatory oversight. The majority of operators are multi-generational families, and all are operating some form of animal-based agriculture, have converted to organic or other value-added products and are located within 1 hour of large metropolitan centres of 2.5 million or more. In addition, three of the study farms are considered small multi-generational family-owned operations. According to United States Department of Agriculture (USDA) classification, a small family owned operation has annual sales of less than $ 250,000. In the United Kingdom, there is no legal definition of a small farm. The Small and Family Farms Association defines a small farm in the United Kingdom as a farm of less than 250 acres. Table 12.1 provides basic farm characteristics for the four case studies.

The following descriptions summarize interviewees, their operations and the pseudonym assigned to each of the four cases. California farmer 1: This was a wife and husband partnership on a 600-acre diversified livestock ranch located in Marin County, Northern California offering grass-fed beef, organic free-range eggs, u-pick apples, seasonal weddings and workshops and a Bed and Breakfast operation. They were interviewed in June 2009. The second California farmer who was interviewed in May 2009 was also from Marin County (California farmer 2). This farm was operated by a wife and husband partnership with daughter and son-in-law on a 800-acre organic dairy farm. In addition to the dairy, they had an on-site winery and tasting room and a Bed and Breakfast tourist business.

The Vermont farmer was interviewed in July 2009. This farm was run by a wife and husband partnership including two sons and their wives on a 220-acre organic dairy farm with a Bed and Breakfast business. The English farm has a wife and husband partnership including daughter and son-in-law on a 92-acre organic dairy farm in the Peak District National Park with a farm Bed and Breakfast, self-catering cottages and on-farm weddings. They were interviewed in September 2009.

Why agritourism?

As noted by Evans and Ilbery (1992), the motivations of farm operators will vary, with small farms' involvement in agritourism activities (e.g. providing

Table 12.1 Characteristics of interviewee's farms

Farm characteristics	*Vermont farmer*	*California farmer 1*	*California farmer 2*	*English farmer*
High land costs	Yes	Yes	Yes	Yes
Farm experience (years)	30	12	50	25
Tourism begun	1984	1997	1997	1990
Next generation involved	Yes	No	Yes	Yes
Land conservation assistance	Vermont Land Trust Department of Agriculture	Marin Ag Land Trust	Marin Ag Land Trust	National Park Authority
Tax relief program	Vermont Dairy Tax Relief Program	Williamson Act	Williamson Act	EU Farm Subsidies

on-farm accommodation services) reflecting a 'survival strategy', whereas operators of larger farms are motivated more by an 'accumulation strategy'. Based on responses, the farms I interviewed seem to be operating from a survival strategy. Diversification allows them to 'bring home a son or daughter' to the farm and help with 'cash flow' or, in some cases, 'keep the farm'.

All three areas, the North Bay area of California, the state of Vermont and the English Peak District, are struggling with the high cost of inputs and low returns on dairy, beef and sheep products. Only 1,042 dairies remain in Vermont, down from 3,063 in 1986, and 102 remain in the Marin and Sonoma county milk shed of the North Bay area. Dairying is also declining in the United Kingdom because of European Union restrictions on individual milk quotas. According to one estimate, a third of all farm operations in the United Kingdom support agritourism activities (Bernardo *et al.* 2004). Agritourism is much less common in the United States, where approximately 52,000 farms earned income from agritourism activities in 2004, representing 2.5 per cent of all farms, according to the USDA (2004).

Interviewees were asked why they began an agritourism business in addition to the original farming operation. The following section describes the motivations of farm operators for initiating agritourism. Although each of these four women operators is motivated by the financial benefits, several other distinct factors emerge. For the Vermont farmer, it was being at home.

> I had to take care of little kids who were 5 and 6, provide a home for my husband, I had to be here. But the farm also needed income. And how could I do that all? Agri-tourism was the key. It was helpful for me because I could be an integral part of the farm business and yet provide that diversification of income and provide a wholesome family experience. But to be able to function as a family and as a home at the same time, you know, it's really been critical for us.

Those with small children on the farm were looking for ways to balance need for income with the desire to remain on the farm. For the California farmer 2, the motivation for beginning agritourism was farm succession.

> We have a great market in the 2.5 million people who visit the area and we are right off the road. The dairy was profitable and was funding the other – but I needed to be where I could manage or be a part of both … and cash flow – and a place to market our wines. So it allowed us to bring our daughter home because we had another income. And she and her husband wanted to operate the dairy.

As farmers in business for only 15 years, the California farmer 1, a farm couple, focuses primarily on making their various farm ventures profitable.

> Our original goals were to make money, preserve and restore the property, and then open the property to visitors. But the bottom line is economics. So, the B&B was first and it was a very nice business and, at the time when our beef business was just getting started and we really didn't have enough animals to make it work and we didn't have enough money to buy more animals, so the B&B worked really well for us.

As in the above quotation, farm diversification was seen by all interviewees as the only strategy that would allow them to continue their business and keep their land. For the English Peak District farm initiating some form of agritourism was a matter of simple survival.

> We moved here (Peak District National Park) 25 years ago in 1984 just as milk quotas were introduced into the European community. So, immediately, our plan to pay our mortgage with increasing our production of milk was scuppered and we had to look for another way of increasing our income. So we brought forward plans that we'd vaguely had about going into tourism. The Peak District National Park is a very well visited tourist destination, and we both enjoyed people's company and we have enough skills to enable us to do all the building work that was necessary. And I'm an entrepreneur …

Diversifying the farm income with agritourism services provided these four families a bridge between farm and family continuity and between enhanced profitability and potential loss of their land.

Linkages between profitability, value-added products and education

The existence of synergy or linkages between operating a farm and operating a tourism enterprise on a farm is an important aspect of the development of agritourism. Early success in farm-based tourism may in fact provide the incentive for farmers to diversify cropping in an effort to meet tourists' demand for speciality foods and value-added products. For example, Cox *et al.* (1995: 210) found in Hawaii that 'tourism encouraged diversified agriculture with the resultant production of high valued, non-traditional crops, such as speciality fruits, coffee, nuts, flowers and nursery products'. However, in a survey of 79 farms in north-eastern England, Sharpley and Vass (2006) identified a widespread desire among farmers to maintain a distinction between the tourism/farming businesses, suggesting that the development of farm tourism enterprises is an employment issue rather than a diversification issue, with the female partner usually responsible for the tourism business.

All these links were apparent in the responses of all four farmers when asked if there were other diversification goals or products that had emerged over time. Tourism may have begun as an opportunity for on-farm employment but changed over time as products were developed (wineries, dairy demonstrations, on-farm meat purchases) as a result of interactions with visitors.

Certainly, these synergies are specific to past and current agricultural production practices and may not apply beyond that industry. On the other hand, new synergies may suggest that today's farm visitors are much more interested in the interaction with the farm and its products than an older generation of visitors. Today's visitors want more than a simple purchase or exchange. They want to experience the farm, getting close up and personal, whether they are buying lamb right off the farm purchased directly from the farmer or paying a farmer to demonstrate and then teach them to milk a cow. Such consumer interest, partially driven by visitors' lost connection to their agricultural roots, provides California farmer 2 with the impetus to expand.

> Actually, the value-added products are now a significant part of our income because the organic dairy conversion, the addition of the winery, and B&B all feed off one another. But originally it was not. We didn't see it then. Farmers are only one and a half percent of the population, the majority of the public doesn't even understand that their food supply comes from farming. We have to do what we can to educate the public. So you have to decide, what's the greater risk, their ignorance or the liability that somebody might fall and sue you, or the liability for an organic farmer that somebody might bring something into the property.

In turn, the interaction with the public changes the farmer. California farmer 2 entertains on-farm activities such as the winery. This project would never have been a consideration in 1997 when I first interviewed this farmer due partially to her fear of lawsuits and lack of visitor experience.

The Vermont farmer talks about their guests and the interplay of sustaining their farm.

> The guest business has been our family living for the last 25 years and it's absolutely critical to the existence of the farm. The cows have to pay the mortgage on the land and buildings. But the cash flow of the guest business allows the farm to continue to function and guests want to participate in farm activities. One of the key components of that when we started it was the – and you can laugh, because this has now become very generally used, but it's kind of the triple bottom line – you know, balancing the economic, the environmental and the social needs. My milking stool has four legs because it's all about the relationships. And if you don't have that, then what's it all about? And so I talk about that balance of, what do we contribute? It's not just economics, it's not just keeping the green spaces, it's not just this, it's not just that. But you really – it's a blend of all the relationships that makes us whole.

Again, each farmer talks about the connections between product development, visitors and the various synergies between them.

For the California farmer 1, the linkages are a mix of income and education.

> We're curious and outgoing people and we like to do a lot of different things. Our neighbours are all involved in some type of livestock operation. We're very interested in the ranch and what it can do, so, you know, subsequently we've gotten into sheep and chickens and pigs and a milk cow, in addition to our grass-fed beef and agritourism activities. But part of it is our interest in local food and animals and our own particular ranch. It's kind of fun to demonstrate that – your land can be healthy and you can also raise food. Because so many people think that agriculture is a big bad guy. I've always been a 'yes, but' person. So, we can say 'Yes, but look at what we're doing here.' It's our way of contributing to the community and helping the community be more viable. You can't exist as an island.

Environmental issues

Acting as environmental stewards is an evolving role. When California farmer 1 restored creeks on the farm in 1996, it was for purely environmental management reasons divorced from any agritourism strategy. Now, this project is incorporated into farm tours and education and is seen as benefiting the community.

Although income was an initial driver, these operations indicate a synergy or blending among the various farm and agritourism operations. This confirms what Mace (2005) and others such as Nickerson *et al.* (2001) and Rilla *et al.* (2010) have already observed that additional opportunities for family employment, community outreach and education plus marketing of farm products work together to enhance any single effort.

Marketing of farm products is especially important on California farms as noted in Chapter 1. Although California farmer 1 highlighted her environmental (creek restoration project) as an evolving component of educating visitors, all the farmers spoke about conservation practices as integral to their success (see Table 12.1).

The Vermont farmer is adamant in her response.

> Yes, absolutely. I think that it's really been an integral part of what I've tried to do to make [visitors] aware that keeping the farm economically viable contributes to the community in many ways. And it's not just the milk product. It's the environmental product that we bring and the societal community part. And, by that, I mean a number of different things. We were founding members of the White River Partnership, which is an environmental stewardship group along the River. We initiated stream bank restoration and excluded our cows with stream setbacks, and encouraged our neighbours to do the same.

The English Peak District farmer living in a National Park is very aware of environmental conservation. However, she has also looked seriously at their farm's energy budget and made substantial improvements.

> I'm already sequestering all this carbon here with good practices. Now you're going to pay me for it. But that's not what stimulates me. What stimulates me is what we can do to change the way that people are. What worries me is that the rural areas in the countryside are left out of the whole renewable energy stuff that's going on, so we have to assess our assets and then use what we've got. We're a south-facing farm, so I thought that solar really has got to have some sort of impact and we'd have an element of solar heating as well. We've got a lot of cow poo here, so can we use that anyway? I came across the plans for the mini bio digester dome, invented in India. They're very small and they're kind of home-made and they do between one cow and 20 cows and they create methane gas for cooking. So I want to see whether we can make that work here and convert our house stoves to work off of it. If it does, it's highly replicable because most country farm properties have oil-fired range cookers and they are just the sort of properties that can't benefit from the sort of massive energy-generating projects that are in urban areas.

California farmer 1 reiterated the importance of her creek project, but this time spoke more in depth about the impact her project had on her farm neighbours and how it has encouraged other conservation practices.

> One of the good changes has been many more of my neighbours have sold their development rights as part of the local agricultural land trust program. So the preservation aspect is strong. The other good thing that's happened is many of my neighbours have done what we originally did, and that is to fence off and plant their creeks. In 1996 we fenced off and planted our portion of the creek, about a half a mile of it, and since then, seven miles of the creek have been fenced off and planted [by my neighbours]. So that's a positive and significant improvement. Organic – our pastures have become organic. You're making me think about this. I haven't thought about this, these changes. And a number of our neighbours have gone organic.

I was impressed with the creativity, persistence and sophistication that all four operators showed in the design and development of the various land conservation practices and projects they shared. One commonality was their passion for the land and their responsibilities as stewards. The English farmer was also aware of the problems of carbon sequestration and the need for renewable energy. Several of the interviewees also spoke about the positive effect their conservation efforts were having on their neighbours and larger community. This suggests that all four women farmers who were interviewed are seen as innovators in their rural communities.

Family transitions at work

The average age of both the UK and American farmers is nearly 60 years, and rural depopulation is a serious issue in some areas of the United States (Bailey and Preston 2007) and England (Williams 1979). When asked about continuing the agritourism venues and the possibility of younger family members managing operations, it was found that each of the interviewees was already in some stage of a generational turnover.

With adult children returning to live and work on the farm, increasing the number of on-farm businesses to support various family members becomes essential. The English farmer expressed this dilemma:

> As we're both getting older, we're finding it much more difficult to manage all the businesses. We were seriously looking at downsizing. How do we stop, you know? Do we get somebody to rent the land? What are our future plans? So we decided that we would look at whether we could create some workers' accommodation here to have help. We talked to the family about it and our daughter and son-in-law said 'you will not bring somebody from Poland or Romania to work for you. We will do it. We will come and live and work here.' And we were absolutely gobsmacked. We were completely stunned. Because we just never had expected that, at least so soon. Our son-in-law wanted to have responsibility for certain aspects, so he bought some organic sheep earlier on in the year and he's managed them. He lambs them and he's now responsible for marketing the lambs to the guests. And now he's taken on board the whole cheese making project.

In this case study, the farmer's worries over farm succession, fear of becoming too frail to farm and exhaustion from keeping 'it all going at once' has a happy ending as family members' needs coalesce. The younger generation has helped to increase diversification through new directions for the farm in terms of organic lamb and cheese-making. The California farmer 2 is at a similar stage to the English farmer.

> We already talked about younger family members taking over the operations. My daughter who is now running the organic dairy with her husband, perhaps my other daughter maybe, who knows, what she may do but that's way in the future. And my niece and nephew were right behind in terms of dairy. And that's kind of it for the family, so really you have a couple of generations sort of lined up that are passionate about remaining in farming.

There are other benefits of having a multigenerational family on the farm. The English farmer expresses the ultimate ideal in this transition:

> You bring people there, you bring income back, you have social care so you don't have children going out to child-minders. You have children, young adults, and then you have old people. So that's exactly what we're going to do here. When we get older, we'll swap with [our children].

The other farm operator in California (California farmer 1) seemed ready to have adult children move back, but timing in this case was not ideal.

> Some of our rancher friends are saying, well, you need to think about bringing your kids into this but our ideal situation would be that one of the kids displayed a real interest in the ranch. So we're – talking about it. And we've asked our kids to be open with one another about their own hopes and dreams and really to talk among one another. I think we could either build another house or – expand on existing quarters and have one of the couples live out here. So what I would really like to see is I would like to see this ranch bring them all together when we're gone, keep them together. And I would like to see the agriculture continue. I think anything that helps any of the ranches survive, is a positive thing and contributes to the viability of the whole community.

California farmer couple 1 are in their late fifties and are beginning to feel the strain of not having adult children on the ranch to assist, whereas the English Peak District farmer is beginning to shift both visitor service and some livestock production responsibilities to adult children. One of the differences may be that the California farm couple 1 have been farming for much less time than the other three farmers who were interviewed (see Table 12.1).

In addition to the challenges of turning the farm over to the younger generation, on-farm housing of returning family members is an important issue. The Vermont farmer and California farmer 2 have suitable accommodation on other agricultural parcels to alleviate housing needs. In the case of the English farmer, increasing on-farm housing involved a complicated planning permit process that ended in a successful appeal. It was especially difficult in this case, as being in the Peak District National Park building is very restricted, and the only possibility is when the new housing is to be used by a member of the farm family. California farmer 1 is contemplating the issue but has not yet made efforts to increase housing and is uncertain whether on-farm housing would be allowable under local zoning regulations. All four cases are in some stage of generational farm transition. On-farm housing is an issue for the majority of family farms in both the United States and United Kingdom, and one that at least these four farms are tackling head-on.

Observations about farm succession and gender

I was struck by the fact that in three of my case studies, it was the daughter (or daughter-in-law) who was returning to manage the dairy or become the herd manager and breeder and who was not simply confined to the visitor service side of a diversified farm.

Strict gender limitations in formerly male-dominated farm careers such as dairy production may be lessening as the next generation of young farmers replace their parents. Although I only asked about the role of young women in the context of younger family members managing the farm (as opposed to a broader context), I did observe that three of the four women farmers whom I interviewed had daughters who are getting involved in both traditional farm production and visitor services.

In the United States, the number of women reporting farming as a livelihood is increasing, with women comprising 14 per cent of principal farm operators and 30 per cent of all farm operators in 2007, a 19 per cent increase over 2002 (U.S. Census of Agriculture, 2007). An article by Johnston (2008) in *The Guardian* newspaper reports that female membership of the British National Farmers' Union has increased by an estimated 50 per cent. The reporter suggests that this is due to women starting smallholdings but indicates that many more are running large, established farms and taking on roles in higher management on farms. There has also been a big increase in women graduates of agricultural colleges in the United Kingdom and in the United States. In the past, such women were limited to jobs in marketing or sales, but now many are starting their own specialist farms, often organic, and becoming large animal veterinarians (Trauger *et al.* 2010).

Farm support and the cost of doing business

This section describes governmental and non-profit policies, practices or programmes and their effects on agritourism efforts. Each farmer was asked about land use or governmental policies that may have helped or hampered their efforts (see Table 12.1). Both California farmers discussed county governing boards and conservation easement programmes as key to their success in remaining in farming.

California farmer 2 comments,

> Our County Board of Supervisors has always been supportive of agriculture and they seem to understand the need for diversification for survival. The farmers' markets all have created a public awareness that, in fact, farming does occur locally. The local organic association, and cheese distributors – and having our ombudsman to help us go through the permit process and so forth, which was very foreign to most of us, because you don't even know what door to go to.

Both farmers sold the development rights on their land and California farmer 1 elaborates here:

> The conservation easement phenomena works because all those people that are supporting the land trust and giving the end of their life legacies, balance of their trusts – because they feel like they are preserving, and they are, a way of life and the land out here that is in active farming and ranching.

Although half of the farmers (both United States and United Kingdom) had experienced permitting challenges, the English farmer expressed frustration with changing standards.

> The black and white policies. So if it said you must do this, then that's what you must do. In the early days of the new environmental economy, quite a lot of work was done to engage planners and get planners on board. But you can't presume that things won't change. One uneducated planner in a position of power changes the equation and can hold up a whole sea change of effort. They simply didn't have an understanding of how a farm operates, how a diverse farm operates. They couldn't understand that the farming operation here that actually maintain this landscape and biodiversity and everything doesn't make money. It should be a presumption that enterprise in the Peak District would be easier on a farm because there's a tacit understanding that enterprise is needed to protect the landscape. So basically we put in an application for three different things: to use our barn for a limited number of weddings, to install a renewable energy project; and to convert a standing stone barn to ancillary accommodation for our kids. We were trying to get a next generation here, working. And we explained it so many different ways, you know, that having ancillary accommodation there meant that these two young people could continue our work here. We could take this landscape forward for the next generation.

According to the study informants, in most cases, local governmental policies have been supportive of agritourism efforts. In addition, conservation easements and an ombudsman who could help with permitting questions were key to farm diversification. In one case, what had been a supportive local authority became less supportive of farming because of a change in personnel.

Regardless of where farms were located, each farmer emphasized the importance of local government officials in understanding and supporting the continued existence of family farming along the urban fringe (interface). The examples they provided included assistance from permitting personnel, for example, ombudsmen or planners, cash from easement programmes and progressive farm-friendly policies.

What will the future hold?

All four operators intend to continue farming and agritourism operations and to pursue the transitional handing over to their children. Here, they describe what they see and hope for in the future, coupled with their fears and concerns.

Both US and UK farmers talk about the urban population and their disconnect with agriculture. The English farmer describes the problem:

> We've got 10 per cent of the population who live in rural areas, while 90 percent live in urban areas. Our normal way of life is so unusual to people who don't live like this, that experiential tourism, sharing your way of life with other people – is huge, almost a forced commitment. If we don't share it in some way – we're not going to be valued anymore. All our politicians and decision makers are urban. It's exactly the same principle in the rest

> of Europe and the United States. Everybody lives in cities. That's really frightening when it's your way of life.

The California farmer 2 is more upbeat, however, perhaps because the rural past is closer in the United States than in the United Kingdom.

> At one time 40 per cent of the people were on farms. They understood where their food came from. And the young people today are so excited about learning how to grow their own food – so I think it's going to improve our standing because they are going to have more empathy and understanding what it takes to have animal husbandry and farming skills.

California farmer 2 talks about a more positive and emerging trend – interest in and about local food production. Although this is a hopeful sign, rural farm areas in the United Kingdom and United States continue to experience depopulation in some areas as cited earlier.

The English farmer describes another rural problem especially important in areas of natural beauty. In a highly urbanized society, the search for a place in the country has been strong for a long time. Although the tax breaks once enjoyed by second home owners have now been much reduced, rich urbanites can still afford to buy country homes, thus driving up rural housing prices and making it impossible for young people to stay in their rural communities. Low-cost housing for local people is a major current problem in English rural areas and there is political pressure for more rural social housing.

> We are losing original farmhouses to industrialists and people who can afford to have a lovely retirement home in the countryside in an old farm, and the villages will continue to be 30 and 40 percent empty second homes. I am working to build into the local development framework process, policies that can take care of farm succession and into the future. The Welsh have accepted the principle of allowing redundant traditional farm buildings on farms to be turned into second dwellings for outsiders although there has been a lot of opposition. Whether it's diversified or agricultural income, it doesn't matter, but it will be there because of the social benefit.

Both California farmers are pushing their local officials to change zoning codes to allow second homes on their agricultural properties to accommodate ageing parents. These progressive policies, if enacted, will soften the transition to the younger farm manager and in some cases provide the tipping point away from land sale.

Encouraging and supporting new and returning farmers is also an important theme as described here by the English farmer:

> It's all about listening and learning and giving people confidence. Nobody else is going to gallop up on a white horse and save their industry or save their family or save their farm, because nobody else cares as much about

> them as they do. So it's actually down to ourselves to do something. And it's giving people that realization, and then the confidence and the power and the enthusiasm to do that.

California farmer 1 adds:

> Once we made the decision to move back to the farm, the lessons began. If you want to be successful in a [farming] community, get involved in the community in every way that you can. Both my husband and I have done that and we've learned so much. Be humble. Be modest. Contribute. Get involved. Don't be a taker. Learn. And then, pull together your own particular skills and interests.

There is a feeling of cautious optimism expressed by the four farmers that California farmer 1 displays:

> I am very hopeful about agriculture in our area. That does not mean that I don't see that it's facing incredible challenges. The most obvious one is our most recent, problem of falling milk prices and milk production, and people being forced to reduce their milk production and even reduce their herds or sell them. Small farms and ranches are challenged all over the United States. On the other hand, our county policies are terrific. There's a lot of talent here. And it can be found among ranch owners, but it's also found among agency people. It's found within the community at large. There are a lot of people who volunteer to keep this community going.

References to young people returning, interest in local food production and farming, counterbalance restrictive land and zoning policies and the strong commitment it takes to keep farming 'going'. These perspectives highlight both the threats and opportunities that await these four farm families as they continue to diversify their operations and bring their children home to farm and live. California farmer 1 sums it up.

> So there are a lot of things that could go wrong and there are a lot of things that could go right. I really feel that any encouragement towards cheese production, small dairying, eco-friendly activities including tourism, the development of model businesses where the land is well taken care of and the animals are well taken care of and the product is healthy for the consumer – you know, developing those kinds of models, I think we just would live forever if we would go in that direction. And I would like to see that. One of my bits of advice would be to take advantage of those treasures that we have here now, don't think they're going to be here forever, because they're not. And don't think it's always going to be this way, because it's not. Change is what's going to happen. And it's either going to be for the better, if you get involved and, if you ignore things, it's going to turn out for the worst.

Conclusion

For these four farms, agritourism has contributed to farm viability by boosting and diversifying cash flow; provided another way for children to return home and enter the farm businesses; and opened a new market and customer base for farm products.

The need for additional on-farm housing was a consistent theme in the four interviews. The desire to house the next generation while keeping 'retirees' on the farm will require local government reconsideration of strict housing limits on these farms. In the US case studies, local government support seemed to be intact, but as the English farmer reported, new building in a National Park often requires lengthy negotiations with the planning authorities even when it is for housing family members. Restrictions on building in National Parks are especially strict with new buildings having to blend in with the existing buildings as planners put conservation of open space for recreation first.

The fact that two of the daughters (daughter-in-law in one case) are entering or returning to traditionally male-dominated agricultural careers is a good sign. According to a study by the Committee on the Future of the Colleges of Agriculture in the Land Grant System, National Research Council (1995) in the United States, women comprised 37.5 per cent of undergraduates and 35 per cent of graduate students in agricultural programmes at US land grant colleges of agriculture in 1993, up by almost 10 per cent from 1982, and this proportion is continuing to grow as women now outnumber men in most higher education institutions.

Small-scale agriculture in the United Kingdom and the United States is still primarily an industry of family farms, passed on to each generation. The more family members willing and able to return can only have a positive impact on family, farm diversification theme and integrated livelihood strategies. Errington and Gasson (1994: 305) summarize the future of family farms by paraphrasing Mark Twain:

> News of the death of the farm family business has been greatly exaggerated. One of its most important advantages over other business forms is its flexibility – it may in fact be a more sustainable business form because it is better able to mutate into forms appropriate to an ever-changing social, economic and political environment.

Certainly, this flexibility and acceptance of change was evident in each case studied. One notable change was the organic transition that occurred in the United States and the United Kingdom. Each operator converted all or some portion of their operations between 1997 and 2009. Although organic production is still small (4 per cent in United Kingdom and <1 per cent in United States), these operators gained both cash and visitor 'goodwill' premiums that made the transition worthwhile. However, Donaldson and Momsen's survey (Chapter 11) showed that visitors rarely came to stay on California farms just for the organic produce. British consumers are more aware of Alternative Food Networks such as organic, Fair Trade and local, so organic farm production is probably more of an attraction for farm visitors there (Harris 2010). Environmental issues are increasingly important on both sides of the Atlantic. Environmental restoration projects completed by each operation parallel programme schemes funded in the past decade by both the EU, UK, and US farm agencies for restoration projects.

It seems that for these operations, the development, evolution and continuance of their agritourism venues have played a key, perhaps in some cases, essential role, in long-term farm viability. All four were still in operation, continuing to diversify and in some cases grow as families derived a mix of social, economic and community benefits from their agritourism business that ultimately kept the farms viable. We know from studies cited including a recent paper by Oredegbe and Fadeyibi (2009) that diversification into farm tourism continues to grow worldwide. Hence, knowledge about what it takes to succeed over the long-term may be valuable. Although these are the experiences of only four operators on the coasts of the United States and in the midlands of England, we can say with certainty that the emphasis on diversification as a survival or succession tactic has been successful to this point. Additional more longitudinal research could illuminate further our understanding of how agritourism is sustaining the farm family, providing opportunity for new on-farm products and assisting in the generational succession within farm families. The parallels between these farm case studies in such different environments support the possibilities of generalizations about the future of agritourism.

References

Bailey, J.M. and Preston, K. (2007) *An Analysis of USDA Farm Program Payments and Rural Development Funding In Low Population Growth Rural Counties*, Lyons, NE: Center for Rural Affairs Rural Research and Analysis Program.

Bernardo, D., Valentin, L. and Leatherman, J. (August 2004) 'Agritourism: If we build it, will they come?', paper presented at the Risk and Profit Conference, Manhattan, KS.

Brown, D.M. and Reeder, R.J. (2007) Farm-Based Recreation, a Statistical Profile, Washington, DC: USDA Economic Research Report No. 53.

Canoves, G. and Villarino, M. (2002) 'Rural tourism gender and cultural conservation in Spain and Portugal', in M. B. Swain and J. Henshall Momsen (eds) *Gender/Tourism/Fun (?)*, New York: Cognizant.

Committee on the Future of the Colleges of Agriculture in the Land Grant System, National Research Council (1995) *Colleges of Agriculture at the Land Grant Universities: A Profile*, Washington, DC, The National Academies Press.

Cox, L., Fox, M. and Bowen, R.L. (1995) 'Does tourism destroy agriculture?', *Annals of Tourism Research*, 22(1): 210–13.

Department for Environment, Food and Rural Affairs (DEFRA) (2004) 'Diversification revenue tops £100 million as farmers' incomes rise again'. Online. Available at: www.defra.gov.uk/news/latest/2004/farm-0104.htm (accessed 22 October 2009).

Errington, A. and Gasson, R. (1994) 'Labour use in the farm family business', *Sociologia Ruralis*, 34(4): 293–307.

Evans, N. and Ilbery, B. (1992) 'Farm-based accommodation and the restructuring of agriculture; evidence from three English counties', *Journal of Rural Studies*, 8(2): 85–96.

George, H. and Rilla, E. (2008) *Agritourism Enterprises on Your Farm or Ranch: Where to Start*, Oakland: University of California, Division of Agriculture and Natural Resources, Publication 8334.

Harris, E.M. (2010) 'East Local? Constructions of Place in Alternative food Politics', *Geography Compass*, 4(4): 355–69.

Hilchey, D. (1993) *Agritourism in New York State*, Ithaca, NY: Farming Alternatives Program, Cornell University.

Ilbery, B., Bowler, B., Clark, G., Crockett, A. and Shaw, A. (1998) 'Farm-based tourism as an alternative farm enterprise: A case study from the northern Pennines, England', *Regional Studies*, 32(4): 355–65.

Jennings, G. and Stehlik, D. (1999) The innovators are women: The development of farm tourism in Central Queensland, Australia, Proceedings of the 1999 International Society of Travel and Tourism Educators Conference, Vancouver, Canada.

Johnston, F. (2008) 'It's dirty, it's noisy – I love my life', *The Guardian*, 5 March 2008. Online. Available at: www.guardian.co.uk/uk/2008/mar/05/ruralaffairs.features (accessed 15 March 2010).

Leff, P. (2009) 'First statewide agritourism survey yields early results', *Small Farm News*, vol. 2 (1 and 11).

McGhee, N.G., Kim, K. and Jennings, G.R. (2007) 'Gender and motivation for agri-tourism entrepreneurship', *Tourism Management*, 28(1): 280–89.

Mace, D. (August 2005) 'Factors motivating agritourism entrepreneurs', paper presented at the Risk and Profit Conference, Manhattan, KS.

Neate, S. (1987) 'The role of tourism in sustaining farm structure and communities on the Isles of Scilly', in M. Bouquet and M. Winter (eds), *Who from their labors rest? Conflict and practice in rural tourism*, Aldershot: Avebury.

Nickerson, N., Black, R.J. and McCool, S.F. (2001). 'Agritourism: Motivations behind farm/ranch business diversification', *Journal of Travel Research*, 40(1): 19–26.

O'Connor, P. (1995) 'Tourism and development in Ballyhoura: Women's business?' *Economic and Social Review*, 26(4): 369–401.

Ollenburg, C. and Buckley, R. (2007) 'Stated economic and social motivations of farm tourism operators', *Journal of Travel Research*, 45(4): 444–52.

Oredegbe, A. and Fadeyibi, I. (June 2009) 'Diversification into farm tourism', paper presented at the International Conference on Regional and Urban Modelling, University of Ottawa, Ontario.

Rilla, E. (1998) 'Unique niches grow in west Marin', *Sustainable Agriculture Technical Review*, 10(2): 3, UC Davis SAREP Program.

Rilla, E., Hardesty, S., Getz, C. and George, H. (2011) 'California agritourism operations are growing', *California Agriculture*, 65(1).

Sharpley, R. and Vass, A. (2006) 'Tourism, farming and diversification: An attitudinal study', *Tourism Management*, 27: 1040–52.

Trauger, A., Sachs, C., Barbercheck, M., Kiernan, N.E., Brasier, K. and Schwartzberg, A. (2010) 'The Object of Extension: Agricultural Education and Authentic Farmers in Pennsylvania', *Sociologia Ruralis*, 30(2): 85–103.

U.S. Census of Agriculture (2007) 'Fact Sheet: Women Farmers', United States Department of Agriculture, National Agricultural Statistics Service. Online. Available at:

www.agcensus.usda.gov/Publications/2007/Online_Highlights/Fact_Sheets/women.pdf (accessed 6 April 2010).

United States Department of Agriculture (USDA) (2004) *Agricultural Resource Management Survey*, Washington, DC: Economic Research Service.

Williams, G. (1979) 'Community development in rural Britain: The organization of social initiative', *Community Development Journal*, 14: 107–14.

13 Visiting winery tasting rooms: venues for education, differentiation and direct marketing

Deborah Che and Astrid Wargenau

Introduction

The agricultural countryside has long been popular with urban visitors who are removed from the farm by a few generations and who hold nostalgic, romanticized views of rural, agricultural areas. European farmers have derived income from renting rooms as well as from directly marketing agricultural produce and local food specialties to such tourists, whereas farmers in peri-urban areas of settler nations offer day activities such as 'U-pick' fruit and vegetables and corn mazes for visitors. Wine tourism, which involves touring wineries and vineyards and tasting wines, is a subset of agricultural tourism. Winery tasting rooms play an important role in augmenting the wine tourism experience by helping to educate consumers about wines, differentiate individual wineries and the region as a wine tourism destination and facilitate direct wine sales. Through a focus on the wineries of southwest Michigan, this study contributes to the literature on tasting rooms in emergent wine regions of the United States. Understanding how tasting rooms function as channels for high-margin cellar door sales can help maintain land in agriculture by increasing returns on investment in areas undergoing rapid residential development.

Winery tasting rooms and wine tourism

Successful wine tourism involves a mix of attributes at the regional level and at the activity place (i.e. the winery and the vineyard). Carmichael (2005) conceptualized the wine tourism experience as in part consisting of core benefits based on tourist motivations, expectations and experiences at the winery or in the region. Core benefits can include relaxation, education or a day in the country. Additionally, facilitating and supporting services and products in the region and at the activity place also figure into the wine tourism experience. At the regional level, facilitating services and products can include the variety of wineries, attraction of winery buildings, rural landscape and transportation network accessibility, whereas supporting services and products can include a wine route, information sources (i.e. wine map brochures), signage, special events, accommodation, catering, shopping and other types of natural and cultural attractions.

Wine routes can strengthen regional attractiveness and destination appeal as they link wineries with complementary resources (Brunori and Rossi 2000; Bruwer 2003; Telfer 2001; Wargenau and Che 2006). This regional attraction mix is built upon offerings at component wineries. At the wineries, facilitating services and products can be shopping, samples, service, tours and main buildings and the vineyard, whereas supporting services and products encompass special events, shopping not related to the main activity, scenic outlooks, wedding receptions and on-site accommodation, restaurants, washrooms and gardens (Carmichael 2005). In the winery tourism experience, tasting rooms are an important part of the regional facilitating services and products at wineries.

In research on the development and marketing of wineries, service quality is consistently stressed. In the study of Getz *et al.* (1999), US (Washington state) survey respondents indicated that friendly and knowledgeable staff, good sign posting, attractive wineries, education/interpretation and special events/functions were the most important factors. In Dodd's (2004) study of visitor influences on cellar door sales and the determinants of wine tourism success in Texas, overall perception of service consisting of attributes such as friendliness, courteousness, knowledge, entertainment, believability and professionalism translated into direct wine purchases. Such direct wine purchases were critical to the success of Texas' small wineries, which produced less than 37,854 litres per year. Winery tasting room visits were their main sales distribution channel. Although these wineries made up 11 per cent of the state total, they accounted for 70 per cent of the tasting room sales. Larger wineries also benefited from visits to winery tasting rooms, in terms of public relations and imparting a positive experience which could lead to future purchases at liquor stores, supermarkets and restaurants (Hall *et al.* 2004). In established wine tourism regions, such as California's Napa Valley, as well as in emergent ones in the US Midwest, tasting room factors including service and the winery environment as well as wine-specific characteristics and price can influence a visitor's attitude towards wine purchases and wine tourism.

Wine in the US heartland

Wines made from native grapes have long been produced in the US Midwest. Wild, native grapevines were the basis for the first plantings of grapevines in 1702 by the French in Detroit. An early American settler, Joseph M. Sterling, planted the first commercial vineyard in Michigan with native Delaware, Concord and Catawba grapes in the mid-1860s. By 1900, Michigan had become one of the top wine-producing states in the United States (Kegerreis and Hathaway 2010).

Although California is by far the dominant wine producer and wine tourism destination in the United States, new Midwest wineries are profitable tourism destinations. Although the extreme continental temperatures in the Plains and Midwestern states generally preclude successful production of European vinifera, newer French-American hybrid grape varieties such as Chambourcin and Seyval

Blanc can survive the cold. On the supply side, farmers can make US$ 1,500/acre (0.4 hectares) in grape and wine production compared with US$ 40/acre for corn[1] and soybeans. On the demand side, homemade wine that reflects the region's character appeals to Midwestern consumers. Together, these forces have reshaped parts of the Iowa landscape, where there are now 70 wineries and more than 700 acres (283.3 hectares) devoted to grape production (Saluny 2006).

Compared with other parts of the US Midwest, certain areas such as Michigan's Lake Michigan shore can grow cold-sensitive European vinifera. The lake's influence extends the January isotherms of −9.4 and −6.67 degrees Celsius over 643.7 kilometres northward (Whitbeck 1920), whereas deep snows associated with downwind, lake effect sites keep the soil from freezing most years, thus preventing root damage. In summer, hot days over 29.4 degrees Celsius are rare. The mild summer days with uniform rains favour fruit production along Lake Michigan's west coast, one of the most important fruit belts in the United States. Although western Michigan wineries initially utilized native grapes that were also used for grape juice, jams and jellies, they shifted their production as American tastes shifted from sweet, dessert-style wines to drier table wines. These wineries began to plant European-American varietals first and then European vinifera. In the mid-1970s, Edward O'Keefe established extensive plantings of vinifera grapes on northwest Michigan's Old Mission Peninsula for his Chateau Grand Traverse winery (Kegerreis and Hathaway 2010). Since 1973, wine grape acreage in Michigan has increased 500 per cent, mainly in vinifera varieties such as Riesling, Chardonnay, Pinot Noir and Cabernet Franc that most American consumers are familiar with (Jones 2006).

Western Michigan has long been and still remains a centre of fruit production and a resulting agritourism destination for those seeking local foods (Che 2006). Western Michigan's moderate temperatures, shore and agricultural lands spurred the development of summer resorts for steamship passengers from Chicago (Kraus 1999; Whitbeck 1920). In the pre-Civil Rights era, some Jewish farm resorts catered to families from Chicago, who sought farm-fresh food, fresh air and Catskills-style entertainment but were excluded from Gentile-only properties (Kraus 1999). More recently, in 2003, the West Coast fruit belt produced 81.6 million kilograms of cherries and 6.6 million kilograms of peaches, making Michigan number one in the world for tart cherry production, number six in the world for sweet cherries and number four in the United States for peaches (*Michigan History* 2006). In total, 1,500 acres (607 hectares) are devoted to wine grapes, most of which grow within 40.2 kilometres of Lake Michigan. Michigan's 50 commercial wineries produce more than 375,000 cases of wine annually, placing it in the top 25 per cent of producers behind California, Washington and Oregon (Michigan Grape and Wine Industry Council 2003). 'More than one million people visit Michigan wineries every year', says Linda Jones, executive director of the Michigan Grape & Wine Industry Council. Michigan has close to 60 wineries, many of which have onsite tasting rooms. Don Coe, managing partner at Black Star Farms Winery in Suttons Bay, says that 60,000 visitors a year come to his tasting room in northern Michigan's Leelanau County.

The vast majority of people buy something, spending on average about US$ 20 to US$ 23 per person (Silfven 2006b). Northwest Michigan is increasingly a culinary tourism destination, named by *The New York Times* as 'Michigan's flavourful vacationland' in recognition of its fruit wines, spirits and innovative regional cuisine (Apple 2003). However, southwest Michigan, with its heritage of fruit production, also draws culinary tourists. These tourists come primarily from northern Indiana, Chicago and Michigan during the peak vacation season. They visit wineries on Fridays, Saturdays and Sundays from Memorial Day (at the end of May) through October.

The focus of this chapter is the wineries of southwest Michigan, including those on the designated Southwest Michigan Wine Trail (Figure 13.1) as well as a specialty producer of 'no-chemical' wine in Kalamazoo. It investigates the role of tasting rooms in the direct marketing of southwest Michigan wines. This chapter looks at how tasting rooms help educate consumers about Michigan wines, how their built environments help differentiate the individual wineries and importantly, how the wine-tasting room experience plays into direct wine sales. We interviewed individuals in charge of the wineries' marketing activities (i.e. marketing directors, members of the marketing departments and the winery owners) over a period from 2003 to 2005. Only when a product or a site specific to a particular winery is discussed are the winery and/or its owners or employees identified by name. At other times, we refer to the wineries by a code letter.

Education at winery tasting rooms

Although southwest Michigan has been a renowned fruit-producing region, it is an emergent wine region. As such, wine-tasting rooms are important in assuring customers of the quality. Sampling before purchasing wines from non-traditional regions can help remove the risk of buying an unknown product (Dodd 2004) as well as augment the perception of service quality. Exposure to and explanation of the region's distinct flavours are part of the winery tasting rooms' educational mission. According to Winery C's winemaker, tasting and education fuel purchases,

> All you got to do is get them through the door, you know. Most of the people, once they come through the door they're gonna buy something and they're gonna be surprised. And the more time they can spend visiting with the winemaker then, you know, that's what it takes.

The visit provides the winemaker with the opportunity to call attention to how writers and judges have favourably compared the region's Rieslings to ones from Germany. Award-winning wines can also be pointed out.

The educational process by knowledgeable staff is particularly important at Peterson and Sons Winery, which is one of the few wineries in the United States and worldwide that does not use any chemicals (i.e. sulphites) in its process. According to wine master Duane Peterson, wine made without chemicals means

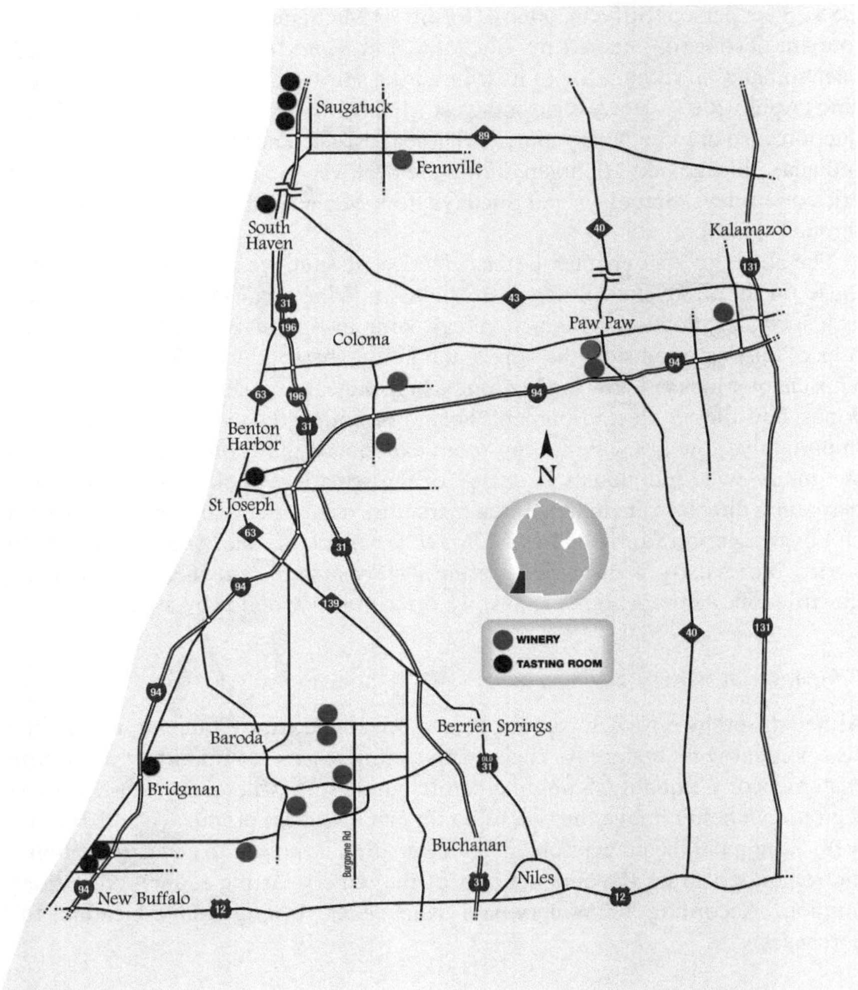

Figure 13.1 Southwest Michigan Wine Trail (adapted from the Southwest Michigan Wine Trail website).

'no allergic reactions such as headaches or sinus problems due to drinking'. Additionally, he informs visitors that grapes are washed prior to crushing to get rid of spray residues, insects, bird droppings and any bad grapes and their vinegar spores, which is particularly important because no chemicals are used. The encounter in the tasting room also provides a way for the wine master to educate customers about proper care of the wine so that they will be able to savour the quality. As his wine is made using only fruit, water, sugar and yeast, several months after bottling carbon dioxide gas can result. In the winery tasting room,

Peterson recommends upright storage out of the sunlight at a constant temperature of 12.8 to 21.1 degrees Celsius. To deal with the carbon dioxide, he recommends pouring a whole bottle of wine slowly into pitchers, swirling or stirring a couple of times and then pouring the wine back into the bottle after the wine is done fizzing (Peterson and Sons Winery 2005). Although production tours are not possible because people would bring in germs against which there is no protection other than the controlled environment in which the wine is made (air filtered to .01 micron or just one step below operating room), interactions with the winemaker and a photo album showing the whole process in the tasting room can illustrate how Peterson's old-fashioned wines with 'the taste of pure fruit' are made.

As knowledge of customers' needs is important in wineries providing excellent service, marketing information can be used to develop long-term relationships with visitors (Dodd 2004). Winery G has a numbered form on its tasting room bar through which customer purchases can be tracked. For customers who may not be able to remember what they bought the last time they were there, the winery can look it up and tell them. Such service at tasting rooms which facilitates relationships is key to the winery experience as its owner explains,

> To me, it's customer service and the quality of your products. If you make it an enjoyable experience for people, they'll tell their friends and they'll come back. So that's really what we try to focus on. We want people to feel like they're part of the family. I mean, this is a family business and it's not always easy to do that when you're very busy, but it's our goal. We try to build a relationship with a person, we don't want to just slam a case in their car and then never see them again. We want them to come back and keep them coming back.

Differentiation through the built environment

In addition to the educational process, the built environment of the tasting rooms is a way to differentiate the southwest Michigan wineries and vary the experience along the Southwest Michigan Wine Trail. This chapter will discuss several of these distinct tasting rooms, which in Carmichael's (2005) model can be considered facilitating services and products at the regional and activity space levels. The Lemon Creek Winery, which originated with a fruit farm owned and operated by the same family since 1855, is rustic and family oriented. In its barn-like setting, it offers affordable wines and non-alcoholic juices. The wine master/owner explained, 'We don't try to go overboard, which is the reason why we are structured the way we are with our prices. We're not gonna charge that super high dollar for high-end wines.' The tasting room reflects the farm-friendly, down-home atmosphere of the operation (Figure 13.2).

In contrast, the Karma Vista Winery is located on a 61-metre hill, which offers spectacular views of fields, orchards and the sand dunes along the lakeshore

Figure 13.2 Lemon Creek Winery Tasting Bar (Photo by Deborah Che)

(supporting services) and a new age atmosphere in the tasting room. The 'karma' part of the name and tasting room, which derives from Joe Herman's philosophy studies at the Marquette University, 'stands for all the good things you do in the vineyard that come back for you years later' (Hoogterp 2004). The winery's banner alludes to good karma in the 'Zen of Merlot Maintenance'. Visitors to the tasting room can experience conversation, an extensive gift selection, the view and a glass of wine with cheese and crackers.

Likewise, the Round Barn Winery offers a tasting room where one can enjoy a glass of wine with cheese, sausages and fruit and a view of the winery's signature round barn. The Amish round barn, which houses a distillery on the ground floor and special events such as weddings on the top floor, shapes the winery's identity and niche (Silfven 2004a; Figure 13.3). This distinctive structure was built in 1911 by farmers who did not want any corners in which bad spirits could hide (Stevenson 2004). Although the farmers were not thinking of bad distilled beverages when they built it, the Round Barn perfectly fits a winery offering good spirits.

Finally, the Contessa Winery's tasting room with mirrors, chandeliers and crystal adds a touch of elegance along the Southwest Michigan Wine Trail. Tasting European varietals and hybrid blends in sheer-rimmed Stolzle crystal stemware is a unique experience. According to wine master Tony Peterson,

Figure 13.3 Round Barn Winery (Photo by Deborah Che)

> Most people don't realize how the delicately shaped bowl of the glass and the rim affect the wine. We have had shocking tastings to prove the difference they make. It's one of our best investments. I have one shot for people to appreciate the wine to its fullest, and you just don't get that in an inexpensive glass.
>
> (Silfven 2006a)

In addition to its regular tastings, Contessa has private ones in which one learns how wine is supposed to be served and stored, what glass is appropriate or how the wine is changed in each different glass. Contessa and the other southwest Michigan wineries are memorable places that add to the wine tourism experience through their built environment and their wine.

Cellar door sales and direct shipping

The wine-tasting rooms provide southwest Michigan wineries with cellar door sales as well as a venue to gain exposure for future direct shipments. Both cellar door sales and direct shipments generate the highest margins. Winery H, which produces 75,000 bottles annually, sells 95 per cent of its wine out of its tasting room as, 'It's rare to have somebody even come in and try four or five wines and not buy a bottle, not buy at least one. But, you still make money'.

Smaller wineries focus on retail rather than wholesale because of the higher margins and the wineries' relatively limited production and staff. Winery B, one of the smaller southwest Michigan wineries, which produces a little over 21,000 bottles of 11 varietal wines, sells only through its tasting room, because,

> If I can sell it all out of my tasting room at full retail, why would I give a wholesale price to put it into the market, which is more hassle for me to go and stock them, deliver it to them, etc. And I am just too small. Down the road sometime I'll have enough fruit to have enough wine where some varietals can be in restaurants and local wine shops.

Given the greater profit, it may make more sense for smaller wineries to focus on cellar door sales by enlarging the tasting room and bar or by adding a tasting room. Winery H noted problems with the size of its tasting room, as on a lot of Saturdays and Sundays 'we can't be any busier here, 'cause you can't get more than three people behind that bar, just doesn't work. And you're stacked three and four deep at the bar and there's nothing else you can do'. Therefore, sales are constrained due to the lack of tasting room capacity. It may also impact perceptions of service quality and the overall wine tourism experience.

Many of the small wineries with tasting rooms are family enterprises, which are often characterized by limited capacity and growth (Getz *et al.* 2004). Limited growth may be due to financial constraints, as a new small family-operated vineyard and winery consisting of 20 acres (8.1 hectares) of wine grapes requires a significant capital investment of more than US$ 700,000. When one figures a cost of US$ 10,000 per acre (0.4 hectares) above the cost of the land to cover the planting and US$ 500,000 for construction of a small tasting room and production facility, winery owners need to borrow significant funds to first get started and to expand later (Silfven 2006a). However, another characteristic of family-owned enterprises is their concern about the loss of distinctive features associated with smallness (Getz *et al.* 2004). Winery C's owner exhibited this when discussing the tasting room,

> Our business plan is to stay small and maximize everything we can do here, you know. If you do another tasting room, people come for the view. You're not selling [just] wine, you are selling all these other things. You know, everybody's sells wine, you're selling you know, the view, the conversation, the chance to relax, the chance to be out in the country all these other things. You couldn't get that in an off-site tasting room, necessarily.

However, having tasting rooms adjoining the vineyards may be key to wine tourism in the countryside of southwest Michigan as compared with urban settings such as Spokane where wineries purchase grapes from the distant Columbia Valley. The situation of smaller Michigan wineries contrasts with that of the largest wineries that have multiple tasting rooms distant from vineyards. The large

Michigan wineries also sell mainly through wholesalers, whereas the smaller ones sell primarily at the tasting rooms, which is a dichotomy between small and large wineries also found in other states and countries (Dodd 2004; Hall *et al.* 2004).

As visits to wineries in Michigan lead to requests for specific wines at liquor stores (Silfven 2004b), they can also spur purchases directly shipped to customers which wineries now have greater opportunities to fulfill. These opportunities follow the 2005 US Supreme Court decision to overturn the Michigan law that did not allow out-of-state wineries to sell directly to Michigan consumers but allowed in-state wineries to do so. Out-of-state and in-state wineries had to be treated equally, which essentially meant that all states that allowed direct shipping to in-state customers via a self-distribution privilege had to give out-of-state wineries the same right (Silfven 2005). Although Michigan consumers could now order directly from Californian wineries, Michigan wineries supported overturning the law because they could now ship directly to customers and members in reciprocal states.[2] Donald Coe, owner of Black Star Farms Winery in Suttons Bay and president of Wine Michigan, a group representing the state's 42 wineries, explained that small wineries had problems with a wholesaler which 'by necessity gives the most attention and seeks the most distribution for brands that are well-established and most recognized by consumers'. Wholesalers did not have much interest in carrying smaller wineries' products. As a result, 'The big wineries get bigger and the little ones just disappear' (*Kalamazoo Gazette 2005*). Thus, the ability to ship directly to consumers enables small wineries to sell to customers in Michigan and in the reciprocal states. According to Linda Jones, executive director of the Michigan Grape and Wine Industry Council, direct shipping on average accounts for at most 5 per cent of winery sales at both large and small Michigan wineries. Not all Michigan wineries have pursued direct shipments to reciprocal states in part because it involves filing paperwork (i.e. for sales and excise taxes) in each of the states the winery wishes to ship to. Direct shipping to Michigan residents actually has gone up in part due to all the publicity over the lawsuit to overturn the Michigan law (Jones 2007). Direct shipping to customers in Michigan and in reciprocal states can grow in the future as Michigan wineries adapt to the changing business climate and as traffic at tasting rooms where most people first sample the offerings of Michigan's smaller wineries increases.

Conclusion

Visits to winery tasting rooms, which have been growing at an annual rate of 10–15 per cent (Jones 2007), are an important part of the wine tourism experience. By sampling and learning about wines, customers can differentiate wineries in the region from one another and from those in other regions. The wine tourism region's built and surrounding natural environment provide facilitating and supporting services. At winery tasting rooms, the wineries can first build relationships and maintain them through future visits and direct shipping (if permitted). Comparing southwest to northwest Michigan, the owner of Winery H said,

> There's a lot more potential in this area. The soils are better here, I mean, the soils, the weather. If we can keep the land in the farming we will be all right, [but] of course, they have the same problem up there.

The problem alluded to is residential development, which is paving over the very unique Fruit Belt farmland. The owner explained the problematic situation,

> In the last two years, I don't think anything sold for under $5,500 an acre (0.4 hectares), which is right up there. And you're talking, I mean, a minimum of like 40 acres (16.2 hectares). So that's right up there. The last farm ground that was even up for sale, didn't even get turned into farm ground again, it got sold for housing development.

Of course, there are two sides to the development. On one hand, the owner of Winery G felt that development could bring more people and customers to the area for cellar door sales. However,

> … the more this area gets developed, the less farm land is going to be available. You know, at our heart we're still a farm and the land here is disappearing … You know, people like to come to this area because it's away from the city and what I don't want to see it become is a city.

The rural landscape functions as a facilitating service for the winery tourism experience. Tasting rooms as a channel for high-margin cellar door sales may help maintain land in agriculture by increasing returns on investment. Like the countryside, the tasting rooms are critical to the regional and activity space levels of the wine tourism experience. Perhaps, Winery F best sums up the importance of service tasting rooms by noting that as a service business,

> We are very, very demanding that we provide pleasant atmosphere for the people that come, so that's an enjoyable experience. And then they leave with a good feeling about the wineries. If you do that, they'll leave with wine and after they get home they will reflect back on the positive experience.

This positive experience can fuel return visits, cellar door sales and orders for directly shipped wine. Winery tasting rooms thus can play an important role in strengthening wine tourism and thus protecting agriculture in a region threatened by sprawling development.

Acknowledgements

The authors thank Cathy Fielding and the Lake Michigan Shore Wine County for use of the map featured in Figure 13.1

Notes

1 However, lately corn prices have escalated due to interest in biofuels, even though extensively grown corn is not as profitable as the small acreages dedicated to wine grapes.

2 Michigan wineries now can ship directly to customers in Alaska, California, Colorado, Hawaii, Idaho, Illinois, Iowa, Minnesota, Missouri, New Mexico, Ohio, Oregon, Washington, Washington, DC, West Virginia and Wisconsin in addition to in-state customers.

References

Apple Jr., R.W. (2003) Up North: Michigan's Flavorful Vacationland. Available at: www.nytimes.com/2003/07/30/dining/30MICH.htm (accessed 31 July 2003).

Brunori, G. and Rossi, A. (2000) 'Synergy and coherence through collective action: Some insights from wine routes in Tuscany', *Sociologia Ruralis*, 40(4): 409–23.

Bruwer, J. (2003) 'South African wine routes: Some perspectives on the wine tourism industry's structural dimensions and wine tourism product', *Tourism Management*, 24(4): 423–35.

Carmichael, B.A. (2005) 'Understanding the wine tourism experience for winery visitors in the Niagara region, Ontario, Canada', *Tourism Geographies*, 7(2): 185–204.

Che, D. (2006) 'Select Michigan: Local food production, food safety, culinary heritage and branding in Michigan agritourism', *Tourism Review International*, 9(4): 349–64.

Dodd, T.H. (2004) 'Influences on cellar door sales and determinants of wine tourism success: Results from Texas wineries', in C.M. Hall, L. Sharples, B. Cambourne and N. Macionis (eds) *Wine Tourism Around the World: Development, Management and Markets*, Oxford, UK: Butterworth Heinemann.

Getz, D., Carlsen, J. and Morrison, A. (2004) *The Family Business in Tourism and Hospitality*, Cambridge, MA: CABI Publishing.

Getz, D., Dowling, R., Carlsen, J. and Anderson, D. (1999) 'Critical success factors for wine tourism', *International Journal of Wine Tourism*, 11(3): 20–41.

Hall, C.M., Johnson, G., Cambourne, B., Macionis, N., Mitchell, R. and Sharples, L. (2004) 'Wine tourism: an introduction', in C.M. Hall, L. Sharples, B. Cambourne and N. Macionis (eds) *Wine Tourism Around the World: Development, Management and Markets*, Oxford, UK: Butterworth Heinemann.

Hoogterp, E. (2004) 'Trail leads travelers to Southwest Michigan wineries; Other popular vineyards are in the Traverse City and Jackson areas', *Kalamazoo Gazette*, 2 May 2004, E1–E2.

Jones, L. (2006) Growth of the Michigan Wine Industry. Available at: www.michiganwines.com/docs/Industry/growth_of_mi_wine_oct06.pdf (accessed 12 March 2010).

Jones, L. (2007) Personal communication, 2 August 2007.

Kalamazoo Gazette. (2005) 'Small wineries fight for survival', 11 October 2005, A1, A4.

Kegerreis, S. and Hathaway, L. (2010) The Early History of the Michigan Wine Industry: Early Settlers of the 1800s to the Passionate Pioneers of the Early 1970s. Available at: www.michiganwines.com/docs/About/early_history_summary.pdf (accessed 12 March 2010).

Kraus, B. (1999) *A Time to Remember: A History of the Jewish Community in South Haven*, Allegan Forest, MI: The Priscilla Press.

Michigan Grape and Wine Industry Council (2003). Michigan Wines. Available at: www.michiganwines.com/index.html (accessed 6 April 2003).

Michigan History (2006) '*The top crops*', 90(3): 13.

Peterson and Sons Winery (2005) 'Michigan's first winery to produce wines with no sulfites added!' Brochure.

Saluny, S. (2006) Iowa Finds Itself Deep in Heart of Wine Country. Available at: www.nytimes.com/2006/11/19/us/19wine1.html?ei=5087%OA (accessed 20 November 2006).

Silfven, S. (2004a) Michigan's Heart of the Vineyard Changing Name. Available at: http://info.detnews.com/wine/columns/silfven/details.cfm?id=210 (accessed 3 June 2007).

Silfven, S. (2004b) Michigan Wine Sales Show Steady Growth. Available at: http://info.detnews.com/wine/columns/silfven/details.cfm?id=262 (accessed 3 June 2007).

Silfven, S. (2005) Michigan Compromise Brings Direct Wine Shipping to the State. Available at: http://info.detnews.com/wine/columns/silfven/details.cfm?id=321 (accessed 3 June 2007).

Silfven, S. (2006a) Starting a New Winery in Michigan Takes Courage and Good Credit. Available at: http://info.detnews.com/wine/columns/silfven/details.cfm?id=332 (accessed 3 July 2007).

Silfven, S. (2006b) It's Not Too Late to Visit a Michigan Winery Tasting Room. Available at: http://forums.detnews.com/wine/columns/silfven/details.cfm?id=369 (accessed 1 June 2007).

Stevenson, J.B. (2004) 'Fruit of the vine', *Grand Rapids*, 41(10): 58–62.

Telfer, D.J. (2001) 'Strategic alliances along the Niagara Wine Route', *Tourism Management*, 22(1): 21–30.

Wargenau, A. and Che, D. (2006) 'Wine tourism development and marketing strategies in southwest Michigan', *International Journal of Wine Marketing*, 18(1): 45–60.

Whitbeck, R.H. (1920) 'The influence of Lake Michigan upon its opposite shores, with comments on the declining use of the lake as a waterway', *Annals of the Association of American Geographers*, 10: 41–55.

14 New forms of tourism in Spain: wine, gastronomic and rural tourism

Gemma Canoves and Raul Suhett de Morais

Introduction

Spain has experienced a continuous expansion of its tourism sector for decades. The most recent data available indicate that it is the second country in the world in number of tourists, with almost 60 million people arriving in 2008 (Escola Universitària de Turisme i Direcció Hotelera [EUTDH] 2009: 15). Spain is now a consolidated, mature tourist destination and its cities are full of travellers from all over the planet. Nowadays, Barcelona, Madrid, Seville, Granada, Santiago de Compostela and Ibiza – just to name a few – are places that everyone has heard of and many are planning to visit.

Spain's tourism became established following the Spanish Civil War during the latter part of the Franco dictatorship as the country's economy resumed growth. Beach tourism was the easiest way of attracting visitors (Ivars 2004), and people from throughout Europe, though primarily from Northern countries, began filling the Spanish beaches during most of the year. Under the slogan 'Spain is different' (used in English even in Spanish documents), the country went from 750,000 tourists in 1950 to more than 24 million tourists in the late 1960s (Lavaina 2006: 7).

Spain occupies the major part of the Iberian Peninsula and is a medium sized country by European standards. However, the Spanish tourism sector was traditionally concentrated in just a few places on the coast. There was no intention to attract tourism inland, as surveys showed that visitors were satisfied with the attractions provided (Instituto Nacional de Estadística [INE] 2008). Nevertheless, after several decades of increasing tourist numbers, Spain realized that it needed to expand beyond traditional sun, sand and sea tourism (Canoves and Villarino 2002). At the end of the 1980s and early 1990s, Spain suffered a tourism crisis and the sector was forced to change its course (Bote 1998; Ivars 2004). The role of tourism in the trade gap fell by half from 1988 until 1992; it was only in 1993 after Barcelona's Olympic Games that the tourism sector regained strength.

At the end of the 1980s, during the transition to democracy, new forms of tourism were introduced in the country. Tourism was growing faster and competition for market share was increasing. Although beach tourism remained the principal

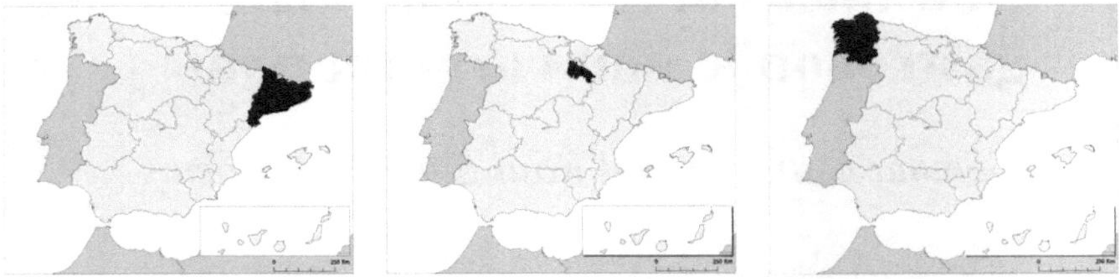

Figure 14.1 The study regions of Catalonia, La Rioja and Galicia, Spain.

drawing point, many Spanish regions were able to add new attractions to their brochures. Iconic buildings were constructed and the XXV Olympic Games were held in Barcelona in 1992, only 10 years after the XII Football World Cup, which had also been held in Spain. The world came to know that Spain was more than nice weather and beautiful beaches.

These changes were more significant in some areas than in others. In this study, we analyse how three different regions (Figure 14.1) in the country faced this new era of tourism. This research has been carried out in various ways in our department through doctoral and master's theses, research groups, field work both in undergraduate and graduate studies and, of course, through our own travel. It is important, however, to clarify that this chapter is based on bibliographical sources to determine what has been said about the changes in Spain's tourism offerings and what such changes have meant to rural areas to define the state of the art of rural tourism in three specific Spanish regions. These regions are representative of the changes that have occurred in the diversification of tourism throughout Spain: Galicia represents the cultural tourism through the pilgrimage of Santiago de Compostela, La Rioja region represents the reinvention of tourism and wine and Catalonia is one clear example of diversification in new rural tourism offerings.

We begin with the least studied of these regions, La Rioja, a small area in northern Spain with a long tradition of wine. Wine is explored as an attraction for a new kind of tourist different from those interested only in the beach. Next, Galicia, in north-western Spain, is a land of complex history, which has successfully established tourist attractions based on its Camino de Santiago (pilgrimage route), rural traditions and the Celtic legacy. The third area of interest is Catalonia, the most industrially developed region in Spain, with many tourism resources, including excursions in the mountains, architecture in rural areas, gastronomic routes and second homes.

This chapter is divided into four parts. First, we consider how Spain reinvented its concept of tourism, highlighting the movement towards the countryside. We analyse the three study regions to understand different trajectories in the development of rural tourism. We then delimit the state of the art of rural tourism in Spain and raise questions for further research, such as a new synergy between tourism and agriculture, the importance for inland areas of rural tourism as economic diversification and the rise of cultural tourism. Finally, we finish with our conclusions on the theme.

Towards a new kind of tourism

According to several scholars (Neal and Garcia-Iglesias 2003; Villarino and Canoves 2000; Yague Perales 2002), tourism in Spain began during the latter years of the Franco dictatorship, which lasted until his death on November 20, 1975. After the Spanish Civil War (1936–1939), the country was destroyed, the population was forced into many years of great privation and the infrastructure became derelict. The country was under an embargo and food was scarce, industry was in ruins, infrastructure was non-existent and no regular transportation existed (Neal and Garcia-Iglesias 2003: 758). Spain was an unlikely vacation destination. Western and northern Europe, the origin of most tourists to Spain (Hermans 1981: 463) were in the midst of World War II. The end of the 1940s and the beginning of the 1950s were years to reconstruct Spain as a tourism destination and to build the economic foundations for tourism demand in Europe.

During the 1960s, Spain's economy started to recover and Europe regained a living standard that allowed people to travel. Spain now became attractive to European travellers as a beach destination. Tourists invaded seeking the good weather of the Mediterranean and of Andalucía's coasts, allowing Spain to become a solid tourism destination. However, tourism levels had stabilized by the 1980s, requiring that Spain rejuvenate its tourism sector leading to the development of rural tourism. Rural tourism is a strategy to combine agriculture and tourism and also benefit rural spaces. Rural tourism was initially promoted by some regional governments, as in the case of Galicia, as a strategy to increase tourism numbers and promote cultural patrimony. In the case of Catalonia, rural tourism began through private initiatives. In general, we assert that Spain adopted the rural strategy to diversifying tourism products by bringing the tourist towards the countryside, offering a wide range of rural and agricultural tourist products.

Spanish towns are filled with magnificent architecture, have strong traditions and hold interesting events. However, Spain was never a big destination for travellers during the days of the Grand Tour taken by the wealthy.[1] Prior to the 1960s, only a few tourists ventured through the country and most of them were experienced travellers. Tourists began paying attention to Spain in the latter half of the twentieth century because of its good weather and eventually because of other attractions: diverse gastronomy and wines, its Catholic and Muslim architectural heritage, the rural landscape, and the idea of the idyllic south.

From 'Spain is different' to a different tourism in Spain

The end of the 1980s was very important to the tourism phenomenon in Spain, for it was the moment of the reinvention of its tourist image. Some interesting changes were occurring, such as the complete transformation of the city of

Barcelona, because of the 1992 Olympic Games (Albet and Riera 1985). Beach tourism had been a successful model for Spain in the early decades of tourism development, but by the 1980s, the need for rejuvenation of the sector to avoid a potential crisis became evident (Butler 1980). Spain had to widen its menu for the tourist.

Given Spain's rich cultural heritage and well-conserved cities, many other possible attractions existed for tourists. Nevertheless, these new destinations would attract a group of tourists looking for meaningful experiences and not just beautiful beaches and good weather. Spain was searching for tourists closer to the individual mass tourist and farther from the organized mass tourist, as described by Cohen (1972), and that required creating new, more elaborate tourism products. Tourism being viewed as the saviour of economic crisis, almost every region of the country took part in the movement to develop new tourism.

Various examples of new tourisms can be observed throughout Spain. Some are based on a rich architectural heritage, such as Seville or Santiago de Compostela; others on various types of routes, such as following the life of El Cid in Andalucía or the wine routes in La Rioja; some others are developed especially for business tourism or sports, such as some cities in the surroundings of Madrid and in Catalonia and so on. In this chapter, however, we will study three specific regions that have developed interesting rural and gastronomic products that allowed them to either avoid stagnation or to start a new tourism activity.

La Rioja, Galicia and Catalonia are the three Spanish regions chosen to illustrate the new forms of rural and gastronomic tourism in Spain emerging during the 1990s. La Rioja provides an example of how a small region with no tradition of tourism is able to attract tourists to its vineyards with well-developed products based on wine and everything related to wine. Galicia, however, had always attracted tourists because of the Camino de Santiago, but it was not until recently that it was converted into a more complex tourist destination combining the Camino with other kinds of rural and gastronomic tourism and profiting from many EU programmes. Finally, in Catalonia, cultural tourism has been well established for at least a century but had to reinvent itself because it needed to rejuvenate to keep its market share.

Bringing the tourist to the countryside

Spain had to diversify its tourism so that it could continue being one of the most visited tourist destinations in the world; therefore, the decision was made to develop products to force the tourist to move to the countryside. This strategy consisted of creating new demands among a different clientele exemplified by Cohen's individual mass tourist (1972), who wants a different experience and understands tourism in a different way.

Until the 1980s, Spain's coastal tourists were of the 'dependable' kind, being 'restrictive in spending discretionary income', 'preferring popular, well-known brands of consumer products' and enjoying 'structure and routine in their relatively non-varying lifestyles' (Plog 2001: 15–16). This type of tourist is easily recognizable in people visiting Spanish beaches in the 1980s, when Spain as a tourist product was reaching the stagnation phase (Butler 1980). To rejuvenate the tourism sector (ibid.), Spain needed to attract the adventurous tourist willing to try new destinations or products (Plog 2001). Plog describes this kind of tourist as willing to 'choose new products shortly after introduction into the marketplace', 'preferring a day filled with varying activities and challenges' and 'spending discretionary income more readily' (ibid. 16–17).

To attract this new type of tourist to Spain, it was necessary to develop new products focusing on the experiential value of the trip (Yague Perales 2002). The Spanish countryside was able to reinvent its image and traditions (Cohen 1988; Brown 1996; Hobsbawn and Ranger 1983; Urry 1995) to meet the needs of the tourist interested in the authentic (Pearce 1990; MacCannell 1999). The entrance of Spain into the European Union was decisive, for significant sums of money were invested in new member states. Much of this was directed to infrastructure, which is a very important part of tourism. However, other investment was more directly aimed at the new tourism beginning in Spain in the countryside, where the tourist would be in contact with the rural and agricultural world (Agarwal 2002; Besteiro Rodríguez 2006; Hoggart and Paniagua 2001; Neal and García-Iglesias 2003; Pérez Fra and López Iglesias 2005; Roquer 2007).

The most well-known example of a EU project was in Galicia, where the Leader (Liaisons Entre Actions de Dévelopment de l'Economie Rurale or French for Links Between Rural Economy Development Actions), Leader II and Leader+ (the second and the third phase of the previous, respectively) and Proder (Programa Operativo para el Desarrollo y la Diversificación Económica en Zonas Rurales or Operative Program for the Development and Diversification in Rural Areas) programmes were implemented. These projects helped Spain (especially Galicia) to change its tourist image and to enlarge its tourist offerings. Although these projects were not designed specifically to improve tourism in the region, their money was deliberately directed to tourism-related ventures (Abad 2003).

The Leader project was created out of the necessity for and the desire to provide sustainable growth in European rural areas. As asserted by Roquer (2007: 124–126), the main characteristics of the programme are territorial focus and the endogenous character, bottom-up methodology, innovation and promotion of new models to be applied in other territories; multi-sector action; and interrelations, cooperation and networking. The first Leader project was implemented from 1989 to 1993, it was followed by Leader II from 1994 to 1999 and Leader+ has been implemented from 2001 forward. Of the three regions studied by this research, Catalonia and Galicia took most profit from these programmes (Pérez Fra and López Iglesias 2005; Roquer 2007). The Proder programme, however, was implemented only in Galicia (of the three studied regions), for it was designed to develop regions with a GDP lower than 75 per cent of the EU average. It was developed by the Spanish government but used a methodology much like Leader projects and shares the same goals.

In the other two regions studied in this chapter, La Rioja and Catalonia, the development of new tourist attractions in the countryside was carried out mostly by the private rather than by the public sector, although some public–private partnerships occurred (Ivars 2004). Tourism is seen as an easy way of complementing an enterprise and sometimes can become the main income earner. Entrepreneurs saw in tourism a new way of selling their products, such as wine in La Rioja; and *haute cuisine*, wine, farm produce and second residency in Catalonia.

The new tourisms in Spain

Despite the tourism potential of the small towns and countryside, they were not taken into account in planning during the first two decades of tourism in Spain. It was much easier to have tourists on the beaches than to have to develop complex products for a much more demanding clientele. Tourism began to develop in the countryside in the late 1980s and the early 1990s, when it was proposed as a secondary activity to help with farm incomes. Until that time, Spanish agriculture was not competitive; thus, tourism activity in rural zones was useful to generate needed income and maintain rural populations. This is why rural tourism began in Catalonia one decade earlier than in the rest of Spain.

The function of tourism as a secondary activity in rural areas did not last long, as it became more attractive to the people involved with it (Busby and Rendle 2000; Canoves *et al.* 2006; Canoves and Villarino 2002). For younger people in particular, tourism is seen as much less demanding work physically than the rural activities of farming. It is common to see older generations working in the fields and youth employed in the tourism sector (Canoves and Villarino 2002; Villarino 1995; Villarino and Canoves 2000). Also, youth are more qualified to work with tourists, because they are generally better educated than their parents, often speaking foreign languages.

The involvement of young people in tourism in small towns of rural Spain is one of the main goals of the many projects carried out (Hoggart and Paniagua 2001). During and after the Spanish Civil War, internal and external emigration was considerable and the big cities received many rural people to work in their incipient factories. Migration drained the countryside, and maintaining the population of the small towns became a real problem. The tourism sector was successful in keeping young people in their native towns instead of going away to Barcelona, Bilbao or Madrid (Neal and Garcia-Iglesias 2003; Paniagua 2002).

La Rioja and its new wine tourism

La Rioja is a small region in northern Spain of just more than 300,000 inhabitants. It covers an area of about 5,000 km^2 (1,947.9 sq mi), which represents 1 per cent of Spanish territory. It has only one university, and research on rural tourism is virtually non-existent. However, tourism – especially the rural kind – is strongly present there.

Although it is the least visited region in Spain in terms of rural tourism (Figure 14.2), during 2008, it was the region in Spain with the biggest increase in tourist

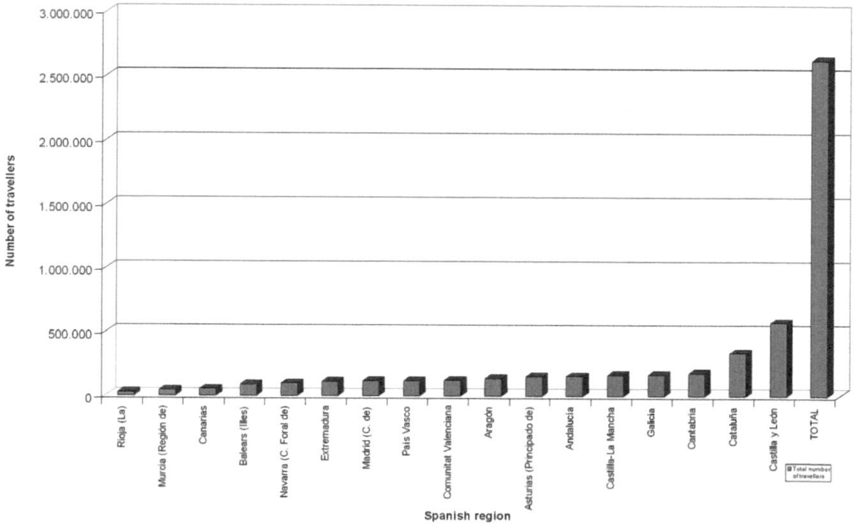

Figure 14.2 Number of rural tourism travellers by Spanish region. Total number of travellers in each of the Spanish regions that had rural tourism as the main motive for his or her trip during 2008. Adapted from INE (2008).

demand (Figure 14.3). La Rioja has a long tradition of rural activities, but tourism is quite new. It did not profit from the first phase of beach tourism in Spain because it is inland. Tourism first emerged in La Rioja less than two decades ago because of interest in the region's wine production (Fuentes *et al.* 2010; Hall *et al.* 2000).

Other wine-producing places around the world were fundamental influences for the establishment of La Rioja's tourist image. Cities such as Porto or Bordeaux were inspirations for the Spanish region to develop a series of products based on wine production, such as cellar visits, wine tasting and *haute cuisine* restaurants, as seen in many tourist guides (La Rioja Guía Viva 2007), magazines and websites (Spain.Info 2009).

La Rioja needs more study, for it is still in the initial stages of tourism development and it needs planning to avoid the negative impacts of tourism. The balance between wine production and tourism must be maintained so that viticulture is not overlooked at the expense of tourism development, which would undermine future tourism prospects.

Galicia and its rural tourism

On the Atlantic coast of Spain, Galicia was always a typical rural region. Its economy was based on farming and fishing, and during and after the Spanish Civil War, it experienced massive emigration. Although Galicia is on the coast, it was not

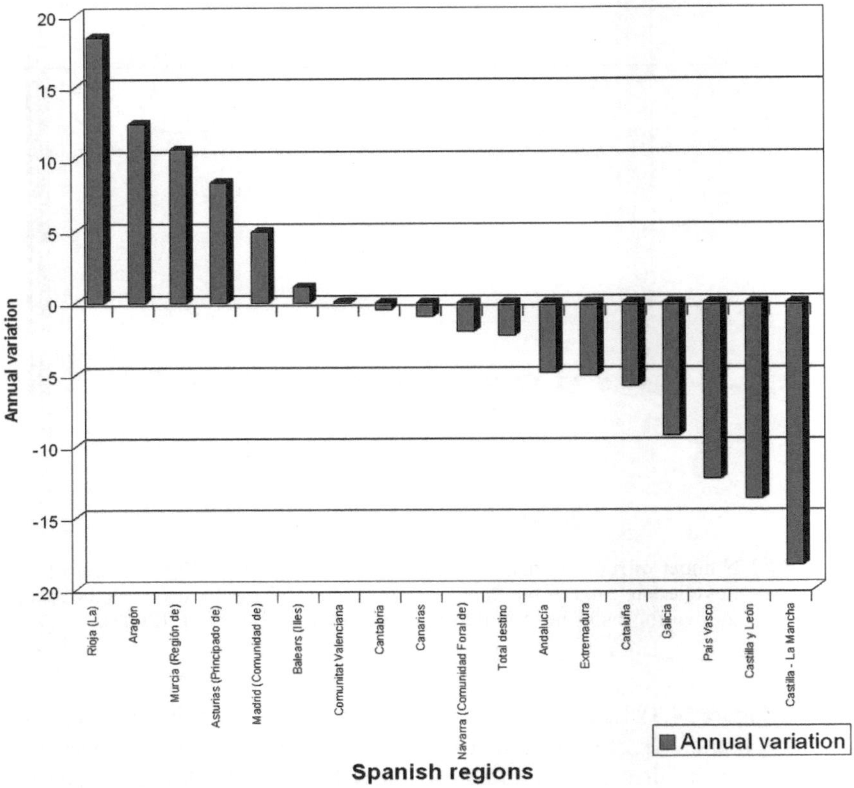

Figure 14.3 Annual variation of tourist demand by Spanish regions. La Rioja had an increase of 18.4 per cent. Adapted from INE (2008).

involved in the first phase of tourism in Spain, mostly because of its wet weather – very different from the year-long dry, sunny weather of the Mediterranean and Andalucía. Galicia is also the most distant location in Spain from other European nations, with poor railway connections hindering development. Even today, Galicia lacks high-speed trains and neither of its main towns have suburban trains.

Galicia, however, has been attracting travellers for a long time because it is the final destination of the Camino de Santiago, one of the most important pilgrimage routes for Catholics (Santos 2006). In this rite, pilgrims cross southern France and the Pyrenees so that they can end their trips in Santiago de Compostela cathedral. This kind of visitor is still present and very important to Galicia's tourist image; nevertheless, new types of tourists are visiting this Spanish region nowadays (Rodríguez 2006).

Until recently, the Camino de Santiago was the only reason to go to Galicia. The pilgrimage was surrounded by religious meanings, such as penitence, sacrifice and personal fulfilment. These feelings are described in some books (such as Paulo Coelho's *The diary of a magus*, 1992) and may have influenced the influx of tourism with motivations beyond the Camino de Santiago itself (Murray and Graham 1997). Santiago de Compostela, Coruña and Vigo are the most important towns in the region and were the entrance gates to tourists. Since Spain joined the European Union, EU projects have influenced the development of a new tourist clientele in Galicia beyond the pilgrim (Smith 1992).

Programmes such as the Leader, Leader+ and Proder were decisive in Galicia's economy. These programmes were not meant to exclusively develop rural tourism, but they invested in it. Analysing the money used from Leader programmes, Pérez Fra and López Iglesias (2005) were able to show how tourism profited from those investments. Depending on the phase of the programme, rural tourism-oriented ventures alternately won the largest share of total investments or represented the largest number of projects funded by the Leader programmes. Sixty-five per cent of all investments were used directly for tourism companies or for other kinds of projects related to tourism, such as infrastructure.

Besides the EU Leader projects, some others were carried out during the 1990s, and it is easy to understand how they assisted Galicia's rural tourism. Abad (2003: 6) shows that rural tourism was virtually non-existent in the early 1990s, but by 2001, almost 4,000 accommodation units were available for this type of tourism. Such development is the result of both investments in tourism offerings (private and public, with regional, national and European levels of financing) and increases in demand (Abad 2003; Martínez and Miguel 2000; Rey and Ramil 2000; Santos 2006). Such growth in demand came after the recent industrialization of Spain, which means that domestic tourists also plan to spend their vacations away from the city, going back to the countryside. Also, the improvement of offerings by rural entrepreneurs and the public sector was a key factor in raising international tourists' travel interests in rural areas.

When it comes to the total number of tourists in Spain, Galicia is of little significance, representing less than 2 per cent (INE 2008). However, in terms of Spanish rural tourism, Galicia is in fourth place after Castilla y León, Catalonia and Cantabria (ibid.) having 6 per cent of the total. Nevertheless, it is important to notice that only Castilla y León and Catalonia have more than 10 per cent of the Spanish rural tourism market.

Of the three case study regions, Galicia is the one with the strongest academic research in rural tourism. In the past 10 years, the experiences of public investments in Galicia were closely observed by university researchers. The county of Pontevedra published a book with a series of papers covering a wide range of subjects about tourism in Galicia and, especially, rural tourism (Hernández *et al.* 2000).

Catalonia and its diverse tourism offerings

Since early times, Catalonia has been the entrance gate for Europeans into Iberia. One fourth of all foreign tourism in Spain occurs in Catalonia (INE 2008; Morais

2008), mostly because of the city of Barcelona's and the Costa Brava's beaches. The Mediterranean spirit has always seemed interesting, especially to Northern Europeans, and Catalonia was one of the most visited regions under the 'Spain is different' slogan, created during the existence of the Ministry of Information and Tourism (1951–1977) (Bayón, 1999: 108; Vega, 2009). It has had a long history of good infrastructure for travellers, especially when compared with other regions of Spain, and many tourist attractions and entrepreneurs who were able to profit from the tourism.

Nevertheless, since the 1992 Olympic Games, tourists are staying longer, giving them time to explore beyond Barcelona and the beaches. Catalonia is the leader in general tourism in the country and the region is also the second largest market for rural tourism in Spain, after Castilla y León, responsible for 12 per cent of Spanish rural tourism. It is also the second region regarding the number of rural tourism accommodation units, with more than 1,500 businesses. Of the 10 most visited provinces, 3 (Lleida, Girona and Barcelona) are Catalan; of the 15 most visited rural zones, 3 (Catalan Pyrenees, Costa Brava and Costa Daurada) are Catalan (INE 2008). These data confirm that Catalonia is more than just Barcelona.

Catalonia's rural tourism is well developed, with a wide range of offerings, from mountain excursions in the Pyrenees to rural houses in the Province of Lleida and from wine tasting in Penedès County to hiking in the Province of Girona. This diversity is a result of early investments in rural tourism, since the 1980s, and tourism and agriculture have had a relationship in Catalonia for more than 30 decades now (Canoves *et al.* 2004; Hermans 1981). It is also interesting to see that rural tourism companies are those with the highest indices of entrepreneurship in the Catalan rural space (López-i-Gelats *et al.* 2009: 610).

The fact that little academic research has been done on Catalonia's rural tourism does not mean that its rural space is ignored. An exhaustive list of studies based on gender (Canoves and Villarino 2000, 2002; García-Ramón *et al.* 1995; García-Ramón *et al.* 1993) and work has been published about the Catalan countryside. One of the reasons for the lack of rural tourism studies is the great importance given to the region's coastline without consideration of the countryside and also because of the low level of public investment in rural tourism in Catalonia. Notably, the Leader programme benefited just a few counties, mainly in the Pyrenees and on the Terres de l'Ebre (Blay 2004; Roquer 2007).

Conclusion

Galicia, La Rioja and Catalonia are three Spanish regions with three different standards of tourism, which we tried to connect through a common pattern: the rural and the agricultural world. Southern European countries have a long tradition regarding the countryside and Spain is no exception. Until the third quarter of the twentieth century, Spain was basically rural and agriculture was still its primary means of income. However, in the late period of Francisco Franco's regime, Spain bet on tourism and job offers moved towards the coastline with investments in the beach tourism model. The beach tourism era promoted the view of 'Spain is different', different from the cold Northern Europe countries; at its beaches, the sun shines, the food is tasty and the hotels are cheap. This model lasted a couple of decades, with tourism increasing and rural areas marching into forgetfulness.

It took a tourism crisis and the entry of Spain into the European Union for rural areas to come back from the oblivion. Programmes such as Leader and Proder were essential to the development of the Spanish countryside, more specifically with tourism as an excuse for development initiatives. Each of our three regions was involved in one way or another with this movement depending on the tourist attractions already established and new products that could be created.

Galicia improved its infrastructure because its tourist attractions were already very much consolidated. It profited from those improvements by luring tourists into staying longer by offering new products in addition to the old ones: rural hotels in addition to the well-known Camino de Santiago. La Rioja entered the tourism world by offering the best product it has: wine. Its *tempranillo* wine was already renowned for its great quality, but the region implemented a whole chain of tourism to offer the tourist the opportunity to know how the wine is made. In Catalonia, the most industrialized region of the peninsula chose to diversify its tourism offerings through the expansion of inland tourism and its high-quality rural hotels, from the Terres de l'Ebre in the South to the Pyrenees, from Barcelona's metropolitan region to Lleida. Catalonia offers the tourist a wide range of facilities and activities.

The main goal of both European and national initiatives to promote the countryside in Spain was to maintain rural populations because internal migration is a big problem for towns of all sizes. Youth were leaving the countryside to live in big cities such as Madrid, Barcelona or Bilbao because of a lack of jobs and the few jobs available were badly paid. Nowadays, with public and private tourism initiatives, even small villages are able to maintain their population and, in some cases, increase it.

Studying the Spanish countryside, its way of life, its history and its relationship with tourism has made us think about how much the rural has been neglected. First, new research groups should dedicate more resources (financial and human) to the study of the relationships between tourism and other rural activities. There are consolidated groups in many universities studying rural and agricultural areas and many others studying tourism; nevertheless, only a few link these two subjects that are part of the life of many people in the countryside.

The second gap we found during the research is the need to cooperate on developing better living standards in the countryside so that people do not need to migrate to the large Spanish cities, yet simultaneously finding ways that the new tourism can protect traditional lifestyles, which are the basis for rural tourism. There are also problems with the manner in which public funds have been invested in rural tourism. Another gap is the improvement in infrastructure, especially needed in Galicia and La Rioja, where public transport is poor, as well as investment in human resources to encourage visitors to spend time in the countryside. We hope that this chapter shows ways to combine rural tourism with protection of the Spanish countryside and maintenance of the way of life of rural people.

Acknowledgements

This chapter was carried out through the I + D + i National Plan's research project 'Turismo, territorio e identidad: procesos de revalorización de espacios y actividades en el medio rural español. Un análisis comparativo de los casos de Cataluña, Galicia y Murcia' financed by Spain's Science and Innovation Ministry. Reference: CSO2009-11793.

Note

1 As defined by Towner (1984: 215), 'the Grand Tour is probably the first extensive tourist movement'. Young aristocrat Europeans (mainly British and Northern Europeans) travelled throughout Europe, especially Italy and France, during the seventeenth and eighteenth centuries, guided by a mentor in a recreational and cultural trip (Adler 1985: 336). It was considered the initiation of the young men by 'the early philosophy of travel as an educational device' (Brodsky-Porge 1981: 171) before they were considered ready to act in society.

References

Abad, P. (2003) 'El turismo rural en Galicia', *Revista Galega de Economía*, 12(2): 1–25.

Adler, J. (1985) 'Youth on the road: reflections on the history of tramping', *Annals of Tourism Research*, 12: 335–54.

Agarwal, S. (2002) 'Restructuring seaside tourism: the resort lifecycle', *Annals of Tourism Research*, 29(1): 25–55.

Albet, A. and Riera, P. (1985) 'Oganització territorial i transformació urbana: de la ciutat industrial a la ciutat espectacle. Una entrevista amb Francesco Indovina' ['Territorial organization and urban transformation: From the industrial city to the spectacle city. An interview with Francesco Indovina'], *Documents d'Analisis Geogràfica*, 33: 109–17.

Bayón, F. (1999) *50 Años del turismo español: un análisis histórico y structural* [50 Years of tourism in Spain: A historical and structural analysis], Madrid: Editorial Centro de Estudios Ramón Areces.

Besteiro Rodríguez, B. (2006) 'El Turismo rural en Galicia: análisis de su evolución en la última década' ['Rural tourism in Galicia: Analysis of evolution in the last decade'], *Cuadernos de Turismo*, 17: 25–49.

Blay, J. (2004) 'L'experiència de Leader-II a les Terres de l'Ebre. Valoració dels principals resultats' ['The Leader-II Experience at the Terres de l'Ebre. Evaluation of the main findings'], in M.A. Pradilla (ed.) *Història i Territori a les comarques de la Diòcesi de Tortosa* [History and territory at the Diocesi de Tortosa counties], Benicarló: Onada edicions.

Bote, V. (1998) 'El desarrollo del turismo en España: cambio de rumbo y oportunidades científicas' ['The development of tourism in Spain: Change of course and scientific opportunities'], *Revista Valenciana d'Estudis Autonòmics*, 25: 29–43.

Brodsky-Porge, E. (1981) 'The grand tour: travel as an educational device 1600–1800', *Annals of Tourism Research*, 8(2): 171–86.

Brown, D. (1996) 'Genuine fakes', in T. Selwyn (ed.) *Tourism image: Myths and myth making in tourism*, Cardiff: Wiley.

Busby, G. and Rendle, S. (2000) 'The transition from tourism on farms to farm tourism', *Tourism Management*, 21: 635–42.

Butler, R. (1980) 'The concept of tourism area cycle of evolution: Implications for management of resources', *Canadian Geographer*, 24(1): 5–12.

Canoves, G. and Villarino, M. (2000) 'Turismo en espacio rural en España: Actrices e imaginario colectivo' ['Tourism in rural space in Spain: Actors and collective imagination'], *Documents d'Anàlisi Geogràfica*, 37: 51–77.

Canoves, G. and Villarino, M. (2002) "Rural tourism gender and cultural conservation in Spain and Portugal", in M.B. Swain and J. Henshall Momsen (eds) *Gender/Tourism/Fun (?)*, New York: Cognizant.

Canoves, G., Villarino, M. and Herrera, L. (2006) 'Políticas públicas, sostenibilidad y turismo rural en España: difícil equilibrio' ['Public policies, sustainability, and rural tourism in Spain: Difficult balance'], *Boletín de la Asociación de Geógrafos Españoles*, 41: 199–217.

Canoves, G., Villarino, M., Priestley, G. and Blanco, A. (2004) 'Rural tourism in Spain: an analysis of recent evolution', *Geoforum*, 35: 755–69.

Coelho, P. (1992) *The diary of a magus*, San Francisco, CA: Harper San Francisco.

Cohen, E. (1972) 'Toward a sociology of international tourism', *Social Research*, 39: 164–82.

Cohen, E. (1988) 'Authenticity and commoditization in tourism', *Annals of Tourism Research*, 15: 371–86.

Escola Universitària de Turisme i Direcció Hotelera (EUTDH) (2009) *Índex UAB d'Activitat Turística: Previsions de demanda turística per a Catalunya i Espanya i estudis especials* [UAB tourism activity index: Tourism demand forecast for catalonia and Spain and special studies], Barcelona: EUTDH.

Fuentes, J.M., Gallego, E., García, A.I. and Ayuga, F. (2010) 'New uses for old traditional farm buildings: the case of the underground wine cellars in Spain', *Land Use Policy*. 27(3): 738–748. Online. Available at: http://www.sciencedirect.com/ (accessed 20 November 2009).

García-Ramón, M.D., Canoves, C. and Valdovinos, N. (1995) 'Farm tourism, gender and the environment in Spain', *Annals of Tourism Research*, 22(2): 267–82.

García-Ramón, M.D., Vilarino, M., Baylina, M. and Canoves, G. (1993) 'Farm women, gender relations and household strategies on the coast of Galicia', *Geoforum*, 24(1): 5–17.

Hall, C., Sharples, L., Cambourne, B. and Macionis, N. (eds) (2000) *Wine tourism around the world: development, management and markets*, Oxford: Butterworth Heineman.

Hermans, D. (1981) 'The encounter of agriculture and tourism: a Catalan Case', *Annals of Tourism Research*, 8(3): 462–79.

Hernández, J., Díaz, J.A., and Patiño, C.A. (eds) (2000) *O turismo en Galicia. Potencialidades endóxenas de desenvolvimento urbano e económico* [Tourism in Galicia: Inside potentialities of urban and economic development], Pontevedra: Deputación Provincial de Pontevedra.

Hobsbawn, E. and Ranger, T. (1983) *The invention of tradition*, Cambridge: Cambridge University Press.

Hoggart, K. and Paniagua, A. (2001), 'The restructuring of rural Spain?', *Journal of Rural Studies*, 17: 63–80.

Instituto Nacional de Estadística (INE) (2008) Instituto Nacional de Estadística. Online. Available at: http://www.ine.es/ (accessed 30 October 2009).

Ivars, J. (2004) 'Tourism planning in Spain: evolution and perspectives', *Annals of Tourism Research*, 31(2): 313–33.

La Rioja – Guía Viva (2007) *La Rioja Guía Viva* [La Rioja Living Guide], Madrid: Anaya Touring Club.

Lavaina, J.C. (ed.) (2006) 'Spain is different: llega la fiebre del turismo', in *Biblioteca el Mundo: El Franquismo Año a Año* [Francoism year to year], Madrid: Unidad Editorial.

López-i-Gelats, F., Tàbara, J.D. and Bartolomé, J. (2009) 'The rural in dispute: Discourses of rurality in the Pyrenees', *Geoforum*, 40: 602–12.

MacCannell, D. (1999) *The tourist*, New York: Schocken Books.

Martínez, F. and Miguel, J. (2000) 'A demanda turística en Galicia. O problema da concentración', *Revista Galega de Economía*, 9(2): 91–108.

Morais, R. (2008) *La imagen de Barcelona a través de las guías turísticas ¿Qué ciudades mostramos?* [The image of Barcelona through tourism guides: what cities do we show?]. Master's Thesis, Cerdanyola del Vallès: Universitat Autònoma de Barcelona.

Murray, M. and Graham, B. (1997) 'Exploring the dialectics of route-based tourism', *Tourism Management*, 18(8): 513–24.

Neal, L. and Garcia-Iglesias, M.C. (2003) 'The economy of Spain without and within the European Union: 1945–2002', *The Quarterly Review of Economics and Finance*, 43: 755–73.

Paniagua, A. (2002) 'Urban–rural migration, tourism entrepreneurs and rural restructuring in Spain', *Tourism Geographies*, 4(4): 349–71.

Pearce, P. (1990) 'Farm tourism in New Zealand: A social situation analysis', *Annals of Tourism Research*, 17: 337–52.

Pérez Fra, M.M. and López Iglesias, E. (2005) 'La contribución del turismo a la diversión de actividades en un espacio rural periférico. Análisis del impacto de la iniciativa Leader en Galicia' ['The contribution of tourism to activity diversification in a peri-rural space'], *Estudios Agrosociales y Pesqueros*, 206: 111–35.

Plog, S. (2001) 'Why destination areas rise and fall in popularity: An update of a Cornell Quarterly Classic', *Cornell Quarterly Classic*, 42: 13–24.

Rey, C. and Ramil, M.C. (2000) 'Estructura do mercado turístico galego' ['Galician tourist market structure'], *Revista Galega de Economía*, 9(1): 293–314.

Rodríguez, B.B. (2006) 'El turismo rural en Galicia. Análisis de su evolución en la última década' ['Rural tourism in Galicia. Analysis of evolution in the last decade'], *Cuardernos de Turismo*, 17: 25–49.

Roquer, S. (2007) 'Els programes comarcals de desenvolupament rural a Catalunya: la iniciativa Leader i el programa Proder' ['The Counties' Programs on rural development at Catalonia: The leader initiative and the Proder Program'], *Treballs de la Societat Catalana de Geografia*, 63: 121–46.

Smith, V. (1992) 'The quest in guest', *Annals of Tourism Research*, 19: 1–17.

Santos, X. (2006). 'El Camino de Santiago: turistas y peregrinos hacia Compostela' ['The Way of Saint James: Tourism and pilgrimage to Campostela'], *Cuardernos de Turismo*, 18: 135–50.

Spain.info (2009) Spain. Online. Available at: http://www.spain.info/ (accessed 30 October 2009).

Towner, J. (1984) 'The grand tour: Sources and methodology for a historical study to tourism', *Tourism Management*, September, 215–22.

Urry, J. (1995) *Consuming places*, London: Routledge.

Vega, C. (2009) *Imatge d'Espanya: Fotografia i Turisme, Conference* [The image of Spain: Photography and tourism], Barcelona: Museu d'Art Contemporani de Barcelona.

Villarino, M. (1995) 'La mujer gallega: un protagonismo permanente' ['The Galician woman: A permanent protagonist'], *El Campo*, 133: 11–24.

Villarino, M. and Canoves, G. (2000) 'Turismo rural en Galicia: sin mujeres imposible' ['Rural tourism in Galicia: Impossible without women'], in M.D. García-Ramon and M. Baylina (eds) *El nuevo papel de las mujeres en el medio rural*, Barcelona: Oikos-Tau.

Yague Perales, R.M. (2002) 'Rural tourism in Spain', *Annals of Tourism Research*, 29(4): 1101–10.

Index